Higher
Chemistry

2001 Exam

2002 Exam

2003 Exam

2004 Exam

2005 Exam

Leckie×Leckie

© Scottish Qualifications Authority

All rights reserved. Copying prohibited. No part of this publication may be reproduced, stored in a retrieval system, or transmitted in any form or by any means, electronic, mechanical, photocopying, recording or otherwise.

First exam published in 2001.
Published by Leckie & Leckie, 8 Whitehill Terrace, St. Andrews, Scotland KY16 8RN tel: 01334 475656 fax: 01334 477392
enquiries@leckieandleckie.co.uk www.leckieandleckie.co.uk
ISBN 1-84372-337-9
A CIP Catalogue record for this book is available from the British Library.
Printed in Scotland by Scotprint.
Leckie & Leckie is a division of Granada Learning Limited, part of ITV plc.

Acknowledgements

Leckie & Leckie is grateful to the copyright holders, as credited at the back of the book, for permission to use their material.
Every effort has been made to trace the copyright holders and to obtain their permission for the use of copyright material.
Leckie & Leckie will gladly receive information enabling them to rectify any error or omission in subsequent editions.

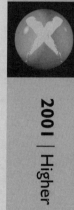

2001 | Higher

[BLANK]

Official SQA Past Papers: Higher Chemistry 2001

FOR OFFICIAL USE

Total Section B

X012/301

NATIONAL QUALIFICATIONS 2001

THURSDAY, 24 MAY 9.00 AM – 11.30 AM

CHEMISTRY HIGHER

Fill in these boxes and read what is printed below.

Full name of centre

Town

Forename(s)

Surname

Date of birth
Day Month Year

Scottish candidate number

Number of seat

Reference may be made to the Chemistry Higher and Advanced Higher Data Booklet (1999 edition).

SECTION A—Part 1 Questions 1–30 and Part 2 Questions 31–35

Instructions for completion of **Part 1** and **Part 2** are given on pages two and seven respectively.

SECTION B

1. All questions should be attempted.

2. The questions may be answered in any order but all answers are to be written in the spaces provided in this answer book, and must be written clearly and legibly in ink.

3. Rough work, if any should be necessary, should be written in this book and then scored through when the fair copy has been written.

4. Additional space for answers and rough work will be found at the end of the book. If further space is required, supplementary sheets may be obtained from the invigilator and should be inserted inside the **front** cover of this book.

5. The size of the space provided for an answer should not be taken as an indication of how much to write. It is not necessary to use all the space.

6. Before leaving the examination room you must give this book to the invigilator. If you do not, you may lose all the marks for this paper.

Scottish Qualifications Authority

SECTION A

PART 1

Check that the answer sheet provided is for Chemistry Higher (Section A).

Fill in the details required on the answer sheet.

In questions 1 to 30 of this part of the paper, an answer is given by indicating the choice A, B, C or D by a stroke made in INK in the appropriate place in Part 1 of the answer sheet—see the sample question below.

For each question there is only ONE correct answer.

Rough working, if required, should be done only on this question paper, or on the rough working sheet provided—**not** on the answer sheet.

At the end of the examination the answer sheet for Section A **must** be placed **inside** this answer book.

This part of the paper is worth 30 marks.

SAMPLE QUESTION

To show that the ink in a ball-pen consists of a mixture of dyes, the method of separation would be

  A fractional distillation

  B chromatography

  C fractional crystallisation

  D filtration.

The correct answer is B—chromatography. A **heavy** vertical line should be drawn joining the two dots in the appropriate box in the column headed **B** as shown **in the example on the answer sheet**.

If, after you have recorded your answer, you decide that you have made an error and wish to make a change, you should cancel the original answer and put a vertical stroke in the box you now consider to be correct. Thus, if you want to change an answer **D** to an answer **B**, your answer sheet would look like this:

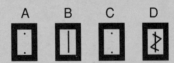

If you want to change back to an answer which has already been scored out, you should **enter a tick (✓)** to the RIGHT of the box of your choice, thus:

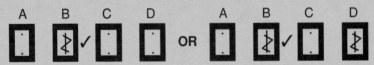

1. A negatively charged particle with electronic configuration 2, 8 could be

   A   a fluoride ion
   B   a sodium atom
   C   an aluminium ion
   D   a neon atom.

2. Which gas would dissolve in water to form an alkali?

   A   HBr
   B NH_3
   C CO_2
   D CH_4

3. $20\,cm^3$ of $0.3\,mol\,l^{-1}$ sodium hydroxide solution can be exactly neutralised by

   A $20\,cm^3$ of $0.3\,mol\,l^{-1}$ sulphuric acid
   B $20\,cm^3$ of $0.6\,mol\,l^{-1}$ sulphuric acid
   C $10\,cm^3$ of $0.6\,mol\,l^{-1}$ sulphuric acid
   D $10\,cm^3$ of $0.3\,mol\,l^{-1}$ sulphuric acid.

4. A mixture of sodium bromide and sodium sulphate is known to contain 5 mol of sodium and 2 mol of bromide ions.

   How many moles of sulphate ions are present?

   A   1·5
   B   2·0
   C   2·5
   D   3·0

5. Excess marble chips (calcium carbonate) were added to $25\,cm^3$ of hydrochloric acid, concentration $2\,mol\,l^{-1}$.

   Which measurement, taken at regular intervals and plotted against time, would give the graph shown below?

   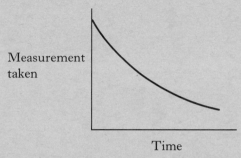

   A   Temperature
   B   Volume of gas produced
   C   pH of solution
   D   Mass of the beaker and contents

6. The following potential energy diagram is for an uncatalysed reaction.

   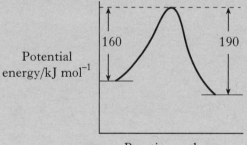

   When a catalyst is used the activation energy of the forward reaction is reduced to $35\,kJ\,mol^{-1}$.

   What is the activation energy of the catalysed reverse reaction, in $kJ\,mol^{-1}$?

   A   35
   B   65
   C   125
   D   155

7. The enthalpy of neutralisation in an acid/alkali reaction is **always** the energy released in

   A   the neutralisation of one mole of acid
   B   the neutralisation of one mole of alkali
   C   the formation of one mole of water
   D   the formation of one mole of salt.

[Turn over

8. Which entry in the table shows the trends in the electronegativity values of the elements in the Periodic Table?

	Across a period	Down a group
A	decrease	decrease
B	decrease	increase
C	increase	decrease
D	increase	increase

9. Which type of structure is found in a fullerene?

   A Ionic lattice

   B Metallic lattice

   C Covalent network

   D Covalent molecular

10. Which type of bonding can be described as intermolecular?

    A Covalent bonding

    B Hydrogen bonding

    C Ionic bonding

    D Metallic bonding

11. An element melts at about room temperature and forms an oxide which reacts with water to form a solution with a pH less than 7.

    Which statement is most likely to be true?

    A The element conducts electricity.

    B The oxide contains covalent bonds.

    C The oxide has a high melting point.

    D The element has a covalent network structure.

12. Which gas occupies the largest volume?

    (Assume all measurements are made at the same temperature and pressure.)

    A 0·32 g of oxygen

    B 0·44 g of carbon dioxide

    C 0·20 g of hydrogen

    D 0·80 g of argon

13. In which reaction is the volume of products less than the volume of reactants?

    A $CH_4(g) + 2O_2(g) \rightarrow CO_2(g) + 2H_2O(g)$

    B $2NH_3(g) \rightarrow N_2(g) + 3H_2(g)$

    C $H_2(g) + Cl_2(g) \rightarrow 2HCl(g)$

    D $2CO(g) + O_2(g) \rightarrow 2CO_2(g)$

14. The Avogadro Constant is the same as the number of

    A molecules in 16 g of oxygen

    B electrons in 1 g of hydrogen

    C atoms in 24 g of carbon

    D ions in 1 litre of sodium chloride solution, concentration $1 \, mol \, l^{-1}$.

15. Biogas is produced under anaerobic conditions by the fermentation of biological materials.

    What is the main constituent of biogas?

    A Butane

    B Ethane

    C Methane

    D Propane

16. Which equation represents a reaction which takes place during reforming?

    A $C_6H_{14} \rightarrow C_6H_6 + 4H_2$

    B $C_4H_8 + H_2 \rightarrow C_4H_{10}$

    C $C_2H_5OH \rightarrow C_2H_4 + H_2O$

    D $C_8H_{18} \rightarrow C_4H_{10} + C_4H_8$

17. Which compound has isomeric forms?

    A Methanol

    B Propane

    C C_2HCl_3

    D $C_2H_4Cl_2$

18. What is the product when one mole of ethyne reacts with one mole of chlorine?

    A 1,1-dichloroethene

    B 1,1-dichloroethane

    C 1,2-dichloroethene

    D 1,2-dichloroethane

19. Which structural formula represents a primary alcohol?

A   CH₃ — CH₂ — CH₂ — C(H)(OH) — CH₃

B   CH₃ — CH₂ — C(H)(OH) — CH₂ — CH₃

C   CH₃ — C(CH₃)(OH) — CH₂ — CH₃

D   CH₃ — C(CH₃)(CH₃) — CH₂ — OH

20. An ester has the structural formula:

CH₃ — CH₂ — C(=O) — O — C(H)(CH₃) — CH₃

On hydrolysis, the ester would produce

A   ethanoic acid and propan-1-ol

B   ethanoic acid and propan-2-ol

C   propanoic acid and propan-1-ol

D   propanoic acid and propan-2-ol.

21. The dehydration of butan-2-ol can produce two isomeric alkenes, but-1-ene and but-2-ene.

Which alkanol can similarly produce, on dehydration, a pair of isomeric alkenes?

A   Propan-2-ol

B   Pentan-3-ol

C   Hexan-3-ol

D   Heptan-4-ol

22. Ozone has an important role in the upper atmosphere because it

A   reflects ultraviolet radiation

B   reflects certain CFCs

C   absorbs ultraviolet radiation

D   absorbs certain CFCs.

23. Which statement can be applied to polymeric esters?

A   They are used for flavourings, perfumes and solvents.

B   They are condensation polymers made by the linking up of amino acids.

C   They are manufactured for use as textile fabrics and resins.

D   They are cross-linked addition polymers.

24. In α-amino acids the amino group is on the carbon atom adjacent to the acid group.

Which of the following is an α-amino acid?

A   CH₃ — CH(CH₂—NH₂) — COOH

B   CH₂(SH) — CH(NH₂) — COOH

C   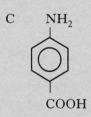 (benzene ring with NH₂ para to COOH)

D   (benzene ring with NH₂ ortho to COOH)

[Turn over

25. Which compound is **not** a raw material in the chemical industry?

A Benzene
B Water
C Iron oxide
D Sodium chloride

26. Which of the following is produced by a batch process?

A Sulphuric acid from sulphur and oxygen
B Aspirin from salicylic acid
C Iron from iron ore
D Ammonia from nitrogen and hydrogen

27.

$$Cl_2(g) + H_2O(\ell) \rightleftharpoons Cl^-(aq) + ClO^-(aq) + 2H^+(aq)$$

The addition of which substance would move the above equilibrium to the right?

A Hydrogen
B Hydrogen chloride
C Sodium chloride
D Sodium hydroxide

28. The concentration of $H^+(aq)$ ions in a solution is 1×10^{-4} mol l^{-1}.

What is the concentration of $OH^-(aq)$ ions, in mol l^{-1}?

A 1×10^{-4}
B 1×10^{-7}
C 1×10^{-10}
D 1×10^{-14}

29. Equal volumes of solutions of ethanoic acid and hydrochloric acid, of equal concentrations, are compared.

In which of the following cases does the ethanoic acid give the higher value?

A pH of solution
B Conductivity of solution
C Rate of reaction with magnesium
D Volume of sodium hydroxide solution neutralised

30. Two 1 g samples of radium and radium oxide both contain the same radioisotope of radium. The intensity of radiation and half-life of the radioisotope in each sample are compared.

Which entry in the table is a correct comparison?

	Intensity of radiation	Half-life
A	same	different
B	same	same
C	different	same
D	different	different

SECTION A

PART 2

In questions 31 to 35 of this part of the paper, an answer is given by circling the appropriate letter (or letters) in the answer grids provided on Part 2 of the answer sheet.

In some questions, two letters are required for full marks.

If more than the correct number of answers is given, marks may be deducted.

In some cases the number of correct responses is NOT identified in the question.

This part of the paper is worth 10 marks.

SAMPLE QUESTION

A CH_4	B H_2	C CO_2
D CO	E C_2H_6	F N_2

(a) Identify the diatomic **compound(s)**.

A	B	C
(D)	E	F

The one correct answer to part (a) is D. This should be circled.

(b) Identify the **two** substances which burn to produce **both** carbon dioxide **and** water.

(A)	B	C
D	(E)	F

As indicated in this question, there are **two** correct answers to part (b). These are A and E. Both answers are circled.

(c) Identify the substance(s) which can **not** be used as a fuel.

A	B	(C)
D	E	(F)

There are **two** correct answers to part (c). These are C and F. Both answers are circled.

If, after you have recorded your answer, you decide that you have made an error and wish to make a change, you should cancel the original answer and circle the answer you now consider to be correct. Thus, in part (a), if you want to change an answer **D** to an answer **A**, your answer sheet would look like this:

(A)	B	C
D̶	E	F

It you want to change back to an answer which has already been scored out, you should enter a tick (✓) in the box of the answer of your choice, thus:

A̶	B	C
✓ D̶	E	F

31. The properties of substances depend on their structures and bonding.

A	B	C
hydrogen	phosphorus	sodium

D	E	F
lithium hydroxide	hydrogen fluoride	hydrogen iodide

(a) Identify the substance with hydrogen bonding between the molecules.

(b) Identify the **two** substances with pure covalent bonding in the molecules.

32. The grid shows the possible effect of temperature change on reaction rate.

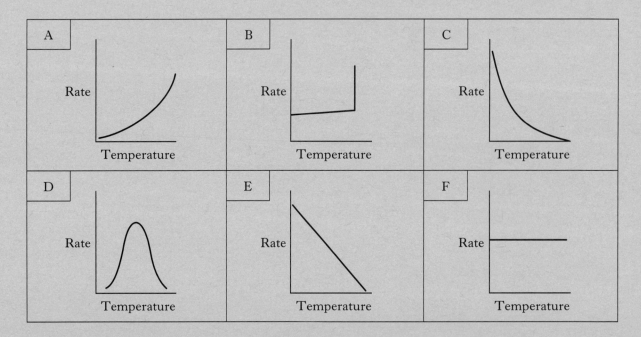

Identify the graph which shows how the rate of reaction varies with temperature in

(a) the decomposition of hydrogen peroxide solution using catalase, an enzyme found in potatoes,

(b) the radioactive decay of phosphorus-32.

33. Propan-2-ol can be prepared from propane as follows.

$$CH_3-CH_2-CH_3 \xrightarrow{\text{Step 1}} CH_3-CH=CH_2 \xrightarrow{\text{Step 2}} CH_3-\underset{\underset{OH}{|}}{CH}-CH_3$$

A	B	C
condensation	cracking	dehydration

D	E	F
hydration	hydrolysis	oxidation

(a) Identify the name of the type of reaction taking place at Step 1.

(b) Identify the name of the type of reaction taking place at Step 2.

34. Identify the statement(s) which can be applied to the role of a catalyst in a reversible reaction.

A	It decreases the enthalpy change for the reaction.
B	It decreases the time required for equilibrium to be established.
C	It alters the equilibrium position.
D	It lowers the activation energy of the backward reaction.
E	It increases the rate of the forward reaction more than the backward reaction.

[Turn over

35. Sodium sulphite is a salt of sulphurous acid, a weak acid.

Identify the statement(s) which can be applied to sodium sulphite.

You may wish to refer to the data booklet.

A	It can be prepared by a precipitation reaction.
B	It can be prepared by the reaction of sulphurous acid with sodium carbonate.
C	In solution, the pH is lower than a solution of sodium sulphate.
D	In redox reactions in solution, the sulphite ion acts as a reducing agent.
E	In redox reactions in solution, the sodium ions are oxidised.

Candidates are reminded that the answer sheet MUST be returned INSIDE this answer book.

SECTION B

1. Technetium-99, which has a long half-life, is produced as a radioactive waste product in nuclear reactors. One way of reducing the danger of this isotope is to change it into technetium-100 by bombardment with particles, as shown by the nuclear equation.

$$^{99}_{43}Tc + X \rightarrow ^{100}_{43}Tc$$

(a) Identify particle **X**.

(b) Technetium-100 decays by beta-emission.
Write a balanced nuclear equation for this reaction.

(c) Technetium-100 has a half-life of 16 s.
If a sample of technetium-100 is left for 48 s, what fraction of the sample would remain?

[Turn over

2. Steam reforming of coal produces a mixture of carbon monoxide and hydrogen.

(a) What name is given to this mixture of carbon monoxide and hydrogen?

(b) This mixture could be used to produce methane, as shown by the following equilibrium.

$$CO(g) + 3H_2(g) \rightleftharpoons CH_4(g) + H_2O(g) \qquad \Delta H = -206 \text{ kJ mol}^{-1}$$

Give **two** reasons why the yield of methane can be increased by cooling the reaction mixture from 400 °C to 80 °C.

3. The effect of temperature changes on reaction rate can be studied using the reaction between an organic acid solution and acidified potassium permanganate solution.

$$5(COOH)_2(aq) + 6H^+(aq) + 2MnO_4^-(aq) \rightarrow 2Mn^{2+}(aq) + 10CO_2(g) + 8H_2O(\ell)$$

The apparatus required is shown in the diagram.

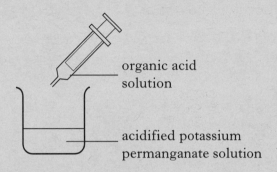

(a) Name the organic acid.

oxalic acid

(b) Describe how the reaction time can be measured.

Time taken for the solution to become colourless

(c) The headings for a set of results are shown below.

Temperature of reaction/°C	Reaction time/s	Reaction rate/ s^{-1}

Complete the headings by entering the correct unit in the third column.

4. A student heated a compound which gave off carbon dioxide and water vapour.

Lumps of calcium chloride were used to absorb the water vapour first, and the carbon dioxide was then collected **in such a way that its volume could be measured**.

(a) Complete the diagram below to show the absorption of water vapour and collection of carbon dioxide.

Label the diagram clearly.

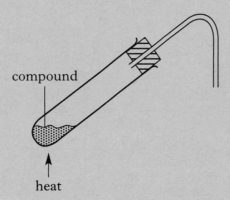

2

(b) The volume of carbon dioxide collected was $240 \, cm^3$.

Calculate the number of molecules in this volume.

(Take the molar volume of carbon dioxide to be 24 litres mol^{-1}.)

1

(3)

5. Vinyl acetate is the monomer for the preparation of polyvinylacetate (PVA) which is widely used in the building industry.

Vinyl acetate has the structural formula:

$$CH_3 - \overset{\overset{\displaystyle O}{\|}}{C} - O - CH = CH_2$$

(a) Draw part of the structure of polyvinylacetate, showing **three** monomer units joined together.

(b) Vinyl acetate and hexane have the same relative formula mass.
Explain why you would expect vinyl acetate to have a higher boiling point than hexane.

[Turn over

6. Aluminium is extracted from bauxite. This ore contains aluminium oxide along with iron(III) oxide and other impurities. The process is shown in the flow diagram.

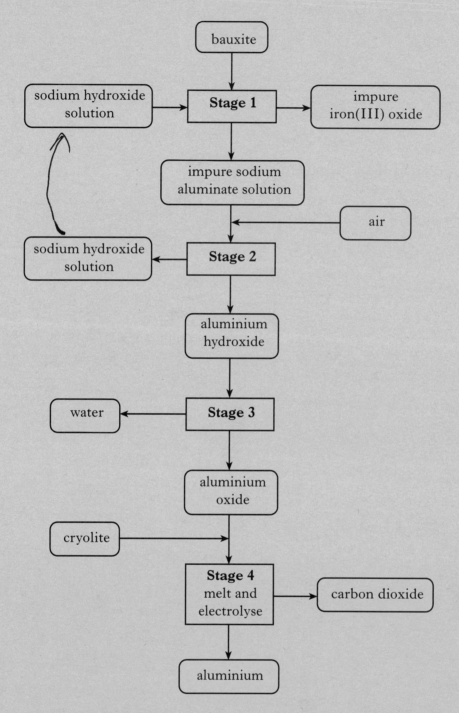

(a) Add an arrow to the flow diagram to show how the process could be made more economical. 1

6. (continued)

   (b) In **Stage 1** of the process, aluminium oxide reacts with sodium hydroxide solution.

   State whether aluminium oxide is behaving as an acidic oxide or as a basic oxide in this reaction.

   acidic oxide

   (c) What type of reaction takes place during **Stage 3**?

   dehydration

   (d) During **Stage 4**, aluminium is manufactured in cells by the electrolysis of aluminium oxide dissolved in molten cryolite.

   What mass of aluminium is produced each hour, if the current passing through the liquid is 180 000 A?

   Show your working clearly.

   $1\,hr = 60 \times 60 = 3{,}600$

   $Q = I \times t$
   $= 180000 \times 3600$
   $= 648{,}000{,}000\ C$

   $3e^- \rightarrow Al$

   $289\,500 \rightarrow 27$
   $648{,}000{,}000 \rightarrow x$
   $x = 60{,}435\,g$
   $x = 60.435\,kg$

   (e) In **Stage 4**, the carbon blocks that are used as positive electrodes must be regularly replaced.

   Suggest a reason for this.

   because they will react with the oxygen and burn away.

7. Potassium hydroxide can be used in experiments to verify Hess's Law. The reactions concerned can be summarised as follows.

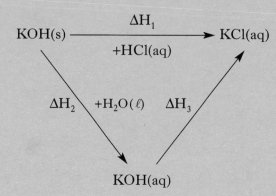

(a) State Hess's Law.

$\Delta H_1 = \Delta H_2 + \Delta H_3$

The empathy change for a chemical is the same no matter what route you take.

(b) Complete the list of measurements that would have to be made in order to calculate ΔH_2.

(i) Mass of potassium hydroxide

(ii) volume of water

(iii) temperature of water at the start

(iv) temperature of water at the end.

(c) What solution must be added to the potassium hydroxide solution in order to calculate ΔH_3?

Hydrochloric acid

8. Calcium hydroxyapatite makes up 95% of tooth enamel.

(a) Tooth decay is caused when tooth enamel is attacked by acid in the mouth.

(i) One of the acids which attacks tooth enamel is 2-hydroxypropanoic acid, which has the molecular formula $C_3H_6O_3$.

Draw a structural formula for this acid.

(ii) Calcium hydroxyapatite reacts with acid in the mouth as shown by the following balanced equation.

$$Ca_{10}(PO_4)_6(OH)_2 + 8H^+ \longrightarrow 10Ca^{2+} + 2H_2O + 6HPO_4^{x-}$$
calcium hydroxyapatite

What is the value of x?

(iii) The pH of a solution in the mouth is 5.

What is the concentration of hydrogen ions, in mol l^{-1}, in this solution?

(b) Tooth enamel also contains a fibrous protein called collagen.

(i) Describe a difference between a fibrous and a globular protein.

(ii) Name the **four** elements present in all proteins.

9. (*a*) Kevlar and Nomex are examples of recently manufactured polymers. Their properties are different because they are made from different monomers.

The diamine monomer used to make Nomex is 1,3-diaminobenzene.

This reacts with the other monomer to form the repeating unit shown.

(i) Draw a structural formula for the other monomer.

(ii) The repeating unit in Kevlar is:

Name the diamine used to make Kevlar.

(*b*) Another recently manufactured polymer is polyvinylcarbazole.

Give the unusual property of polyvinylcarbazole which makes it suitable for use in photocopiers.

10. (a) Propanone and propanal both contain the same functional group.

(i) Name this functional group.

(ii) The diagram shows how to distinguish between propanone and propanal.

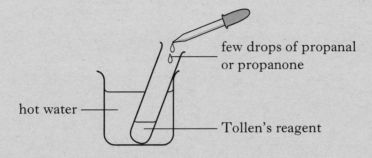

Explain what is observed in the reaction between propanal and Tollen's reagent.

(b) The equation for the enthalpy of formation of propanone is:

$$3C(s) + 3H_2(g) + \tfrac{1}{2}O_2(g) \longrightarrow C_3H_6O(\ell)$$

Use the following information on enthalpies of combustion to calculate the enthalpy of formation of propanone.

$$C(s) + O_2(g) \to CO_2(g) \quad \Delta H = -394 \text{ kJ mol}^{-1}$$
$$H_2(g) + \tfrac{1}{2}O_2(g) \to H_2O(\ell) \quad \Delta H = -286 \text{ kJ mol}^{-1}$$
$$C_3H_6O(\ell) + 4O_2(g) \to 3CO_2(g) + 3H_2O(\ell) \quad \Delta H = -1804 \text{ kJ mol}^{-1}$$

Show your working clearly.

11. (a) Mordenite is a porous, crystalline material with a surface area of over $500\,m^2\,g^{-1}$.

It is used in an isomerisation reaction, part of a sequence which converts pentane into 2-methylbutane for blending into petrol.

pentane ⟶ pent-2-ene —isomerisation⟶ 2-methylbut-2-ene ⟶ 2-methylbutane

(i) Draw a structural formula for 2-methylbut-2-ene.

(ii) What role does mordenite play in the isomerisation reaction?

(iii) Why is 2-methylbutane a more suitable component than pentane when used in unleaded petrol?

(b) Mordenite consists mainly of silicon dioxide.
Name the structure and type of bonding in silicon dioxide.

12. On crossing the Periodic Table, there are trends in the sizes of atoms and ions.

(a) Why is the atomic size of chlorine less than that of sodium?

(b)

Ion	Ionic radius/pm
Si^{4+}	42
P^{3-}	198

Why is there a large increase in ionic radius on going from Si^{4+} to P^{3-}?

13. Ammonium chloride (NH_4Cl) is soluble in water.

(a) How does the pH of a solution of ammonium chloride compare with the pH of water?

1

(b) A student dissolved 10·0 g of ammonium chloride in 200 cm³ of water and found that the temperature of the solution fell from 23·2 °C to 19·8 °C.
Calculate the enthalpy of solution of ammonium chloride.
Show your working clearly.

2
(3)

DO NOT WRITE IN THIS MARGIN

Marks

14. (*a*) Some carbon monoxide detectors contain crystals of hydrated palladium(II) chloride. These form palladium in a redox reaction if exposed to carbon monoxide.

$$CO(g) + PdCl_2.2H_2O(s) \longrightarrow CO_2(g) + Pd(s) + 2HCl(g) + H_2O(\ell)$$

Write the ion-electron equation for the reduction step in this reaction. **1**

(*b*) Another type of detector uses an electrochemical method to detect carbon monoxide.

At the positive electrode:

$$CO(g) + H_2O(\ell) \longrightarrow CO_2(g) + 2H^+(aq) + 2e^-$$

At the negative electrode:

$$O_2(g) + 4H^+(aq) + 4e^- \longrightarrow 2H_2O(\ell)$$

Combine the two ion-electron equations to give the overall redox equation. **1**

(2)

[Turn over

15. Sugars, such as glucose, are often used as sweeteners in soft drinks.

The glucose content of a soft drink can be estimated by titration against a standardised solution of Benedict's solution. The copper(II) ions in Benedict's solution react with glucose as shown.

$$C_6H_{12}O_6(aq) + 2Cu^{2+}(aq) + 2H_2O(\ell) \longrightarrow Cu_2O(s) + 4H^+(aq) + C_6H_{12}O_7(aq)$$

(a) What change in the ratio of atoms present indicates that the conversion of glucose into the compound with molecular formula $C_6H_{12}O_7$ is an example of oxidation?

(b) In one experiment, $25.0\,\text{cm}^3$ volumes of a soft drink were titrated with Benedict's solution in which the concentration of copper(II) ions was $0.500\,\text{mol}\,l^{-1}$. The following results were obtained.

Titration	Volume of Benedict's solution/cm³
1	18·0
2	17·1
3	17·3

Average volume of Benedict's solution used = $17.2\,\text{cm}^3$.

(i) Why was the first titration result not used in calculating the average volume of Benedict's solution?

15. *(b)* **(continued)**

(ii) Calculate the concentration of glucose in the soft drink, in mol l^{-1}.
Show your working clearly.

(c) In some soft drinks, sucrose is used instead of glucose.

Why can the sucrose concentration of a soft drink **not** be estimated by this method?

16. In experiments with four different gases, a syringe was held vertically as shown with the weight of the syringe piston applying a downward pressure on the gas. The times taken for 60 cm³ of helium, methane, carbon dioxide and butane to escape through the pinhole were measured and the graph shows the results plotted against relative formula mass.

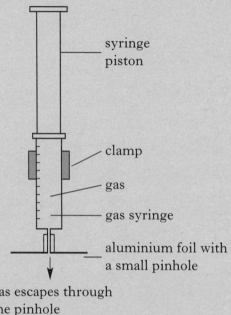

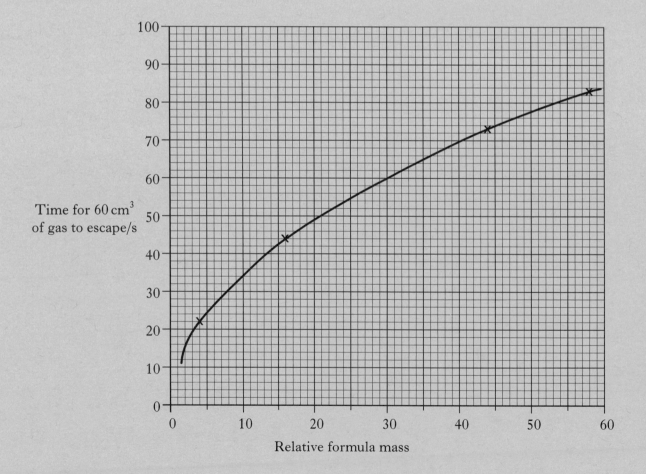

16. (continued)

(a) Calculate the average rate of escape from the syringe of 60 cm³ of methane, in cm³s⁻¹.

(b) Name a hydrocarbon gas which would take 56 s to escape.

(c) The error in a measurement decreases as the actual size of the measurement increases.
Suggest **one** way of reducing the error in each of the time measurements.

[Turn over for Question 17 on *Page thirty*]

17. Biodiesel is a mixture of esters which can be made by heating rapeseed oil with methanol in the presence of a catalyst.

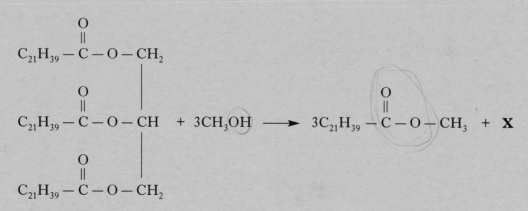

a triglyceride in rapeseed oil   methanol   a component of biodiesel

(a) Name compound **X**.

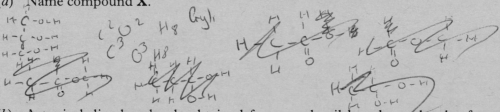

(b) A typical diesel molecule obtained from crude oil has the molecular formula $C_{16}H_{34}$ (hexadecane).

Other than the ester group, name a functional group present in biodiesel molecules which is **not** present in hexadecane.

(c) Vegetable oils like rapeseed oil are converted into fats for use in the food industry.

What name is given to this process?

hydrogenation

[END OF QUESTION PAPER]

2002 | Higher

[BLANK]

FOR OFFICIAL USE

Total Section B

X012/301

NATIONAL QUALIFICATIONS 2002	TUESDAY, 4 JUNE 9.00 AM – 11.30 AM	CHEMISTRY HIGHER

Fill in these boxes and read what is printed below.

Full name of centre

Town

Forename(s)

Surname

Date of birth
Day Month Year   Scottish candidate number   Number of seat

Reference may be made to the Chemistry Higher and Advanced Higher Data Booklet (1999 edition).

SECTION A—Part 1 Questions 1–30 and Part 2 Questions 31–35

Instructions for completion of **Part 1** and **Part 2** are given on pages two and eight respectively.

SECTION B

1. All questions should be attempted.

2. The questions may be answered in any order but all answers are to be written in the spaces provided in this answer book, and must be written clearly and legibly in ink.

3. Rough work, if any should be necessary, should be written in this book and then scored through when the fair copy has been written.

4. Additional space for answers and rough work will be found at the end of the book. If further space is required, supplementary sheets may be obtained from the invigilator and should be inserted inside the **front** cover of this book.

5. The size of the space provided for an answer should not be taken as an indication of how much to write. It is not necessary to use all the space.

6. Before leaving the examination room you must give this book to the invigilator. If you do not, you may lose all the marks for this paper.

SCOTTISH QUALIFICATIONS AUTHORITY

SECTION A

PART 1

1. Check that the answer sheet provided is for Chemistry Higher (Section A).

2. Fill in the details required on the answer sheet.

3. **In questions 1 to 30 of this part of the paper, an answer is given by indicating the choice A, B, C or D by a stroke made in INK in the appropriate place in Part 1 of the answer sheet—see the sample question below.**

4. **For each question there is only ONE correct answer.**

5. Rough working, if required, should be done only on this question paper, or on the rough working sheet provided—**not** on the answer sheet.

6. At the end of the examination the answer sheet for Section A **must** be placed **inside** the front cover of this answer book.

This part of the paper is worth 30 marks.

SAMPLE QUESTION

To show that the ink in a ball-pen consists of a mixture of dyes, the method of separation would be

    A fractional distillation

    B chromatography

    C fractional crystallisation

    D filtration.

The correct answer is B—chromatography. A **heavy** vertical line should be drawn joining the two dots in the appropriate box in the column headed **B** as shown **in the example on the answer sheet**.

If, after you have recorded your answer, you decide that you have made an error and wish to make a change, you should cancel the original answer and put a vertical stroke in the box you now consider to be correct. Thus, if you want to change an answer **D** to an answer **B**, your answer sheet would look like this:

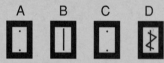

If you want to change back to an answer which has already been scored out, you should **enter a tick (✓)** to the RIGHT of the box of your choice, thus:

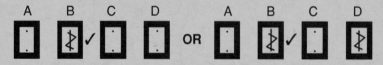

1. Which covalent gas dissolves in water to form an alkali?

   A HCl
   B CH_4
   C SO_2
   D NH_3

2. When copper is added to a solution containing zinc nitrate and silver nitrate

   A deposits of both zinc and silver form
   B a deposit of zinc forms
   C a deposit of silver forms
   D no new deposit forms.

3. Hydrochloric acid reacts with magnesium according to the following equation.

   $$Mg(s) + 2H^+(aq) \rightarrow Mg^{2+}(aq) + H_2(g)$$

   What volume of 4 mol l^{-1} hydrochloric acid reacts with 0·1 mol of magnesium?

   A 25 cm^3
   B 50 cm^3
   C 100 cm^3
   D 200 cm^3

4. Two identical samples of zinc were added to an excess of two solutions of sulphuric acid, concentrations 2 mol l^{-1} and 1 mol l^{-1} respectively.

   Which of the following would have been the same for the two samples?

   A The total mass lost
   B The total time for the reaction
   C The initial reaction rate
   D The average rate of evolution of gas

5.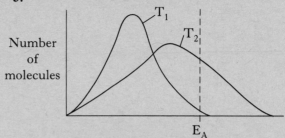

   Which of the following is the correct interpretation of the above energy distribution diagram for a reaction as the temperature **decreases** from T_2 to T_1?

	Activation energy (E_A)	Number of successful collisions
A	remains the same	increases
B	decreases	decreases
C	decreases	increases
D	remains the same	decreases

6. The potential energy diagram for the reaction

   $$CO(g) + NO_2(g) \rightarrow CO_2(g) + NO(g)$$

   is shown.

   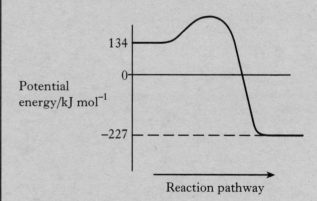

   ΔH, in kJ mol^{-1}, for the forward reaction is

   A −361
   B −227
   C −93
   D +361.

   [Turn over

7. Which type of bond is broken when ice is melted?

   A Ionic

   B Polar covalent

   C Hydrogen

   D Non-polar covalent

8. The shapes of some common molecules are shown below and each contains at least one polar bond.

   Which molecule is non-polar?

   A H—Cl

   B H—O—H (bent)

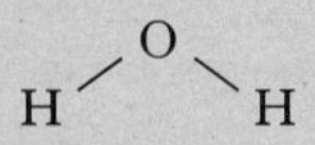

   C O=C=O

   D 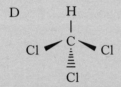 CHCl$_3$ tetrahedral

9. A metal (melting point 843 °C, density 1·54 g cm^{-3}) was obtained by electrolysis of its chloride (melting point 772 °C, density 2·15 g cm^{-3}) at 780 °C.

   During the electrolysis, how would the metal occur?

   A As a solid on the surface of the electrolyte

   B As a liquid on the surface of the electrolyte

   C As a solid at the bottom of the electrolyte

   D As a liquid at the bottom of the electrolyte

10. Which of the following contains the **largest** number of molecules?

    A 0·10 g of hydrogen gas

    B 0·17 g of ammonia gas

    C 0·32 g of methane gas

    D 0·35 g of chlorine gas

11. The equation for the complete combustion of propane is:

    $$C_3H_8(g) + 5O_2(g) \rightarrow 3CO_2(g) + 4H_2O(\ell)$$

    50 cm^3 of propane is mixed with 500 cm^3 of oxygen and the mixture is ignited.

    What is the volume of the resulting gas mixture?

    (All volumes are measured at the same temperature and pressure.)

    A 150 cm^3

    B 300 cm^3

    C 400 cm^3

    D 700 cm^3

12. It is now known that protons and neutrons are made up of smaller particles called quarks.

    Each proton and each neutron contains 3 quarks.

    What is the approximate number of quarks in 1 g of carbon-12?

    A 6×10^{23}

    B 9×10^{23}

    C $1·8 \times 10^{24}$

    D $2·16 \times 10^{25}$

13. Which pollutant, produced during internal combustion in a car engine, is **not** the result of incomplete combustion?

    A Nitrogen dioxide

    B Hydrocarbons

    C Carbon

    D Carbon monoxide

14. Which equation represents an industrial reforming process?

    A $CH_3(CH_2)_6CH_3 \rightarrow CH_3(CH_2)_4CH_3 + CH_2=CH_2$

    B $CH_3(CH_2)_6CH_3 \rightarrow CH_3C(CH_3)_2CH_2CH(CH_3)_2$

    C $CH_3(CH_2)_6CH_2OH \rightarrow CH_3(CH_2)_5CH=CH_2 + H_2O$

    D $4CH_2=CH_2 \rightarrow -(CH_2CH_2)_4-$

15. Which of the following is an isomer of hexanal?

A 2-methylbutanal

B 3-methylpentan-2-one

C 2,2-dimethylbutan-1-ol

D 3-ethylpentanal

16. An ester is prepared from methanoic acid and ethanol.

Which of the following is the full structural formula for the ester produced?

A
```
        H   H       O
        |   |       ||
    H — C — C — O — C — H
        |   |
        H   H
```

B
```
        H   O       H
        |   ||      |
    H — C — C — O — C — H
        |           |
        H           H
```

C
```
        H   H       H
        |   |       |
    H — C — C — O — C — H
        |   |       |
        H   H       H
```

D
```
        H   H   O
        |   |   ||
    H — C — C — C — O — H
        |   |
        H   H
```

17. Which statement about benzene is correct?

A Benzene is an isomer of cyclohexane.

B Benzene reacts with bromine solution as if it is unsaturated.

C The ratio of carbon to hydrogen atoms in benzene is the same as in ethyne.

D Benzene undergoes addition reactions more readily than hexene.

18. Which reaction can be classified as reduction?

A $CH_3CH_2OH \rightarrow CH_3COOH$

B $CH_3CH(OH)CH_3 \rightarrow CH_3COCH_3$

C $CH_3CH_2COCH_3 \rightarrow CH_3CH_2CH(OH)CH_3$

D $CH_3CH_2CHO \rightarrow CH_3CH_2COOH$

19. Part of a polymer molecule is represented below.

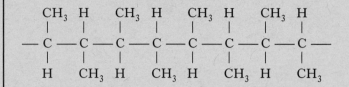

The monomer which gives rise to this polymer is

A but-2-ene

B but-1-ene

C methylpropene

D buta-1,3-diene.

20. Which mixture of gases is known as synthesis gas?

A Methane and oxygen

B Carbon monoxide and oxygen

C Carbon dioxide and hydrogen

D Carbon monoxide and hydrogen

21. Some recently developed polymers have unusual properties.

Which polymer is soluble in water?

A Poly(ethyne)

B Poly(ethenol)

C Biopol

D Kevlar

[Turn over

22. When two amino acids condense together, water is eliminated and a peptide link is formed.

Which of the following represents this process?

A H R₁ O H R₂ O
  \ | // | | //
   N—C—C   N—C—C
  / | \ / | \
 H H OH H H OH

B H R₁ O R₂ O
  \ | // | //
   N—C—C H—C—C
  / | \ | \
 H H OH N OH
           / \
          H   H

C H R₁ O H R₂ O
  \ | // | | //
   N—C—C   N—C—C
  / | \ / | \
 H H OH H H OH

D H R₁ O O R₂ H
  \ | // // | /
   N—C—C   C—C—N
  / | \ / | \
 H H OH H—O H H

23. Consider the reaction pathway shown.

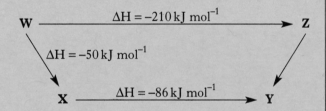

According to Hess' Law, the ΔH value, in kJ mol⁻¹, for reaction **Z** to **Y** is

A +74
B −74
C +346
D −346.

24. Which of the following is likely to apply to the use of a catalyst in a chemical reaction?

	Position of equilibrium	Effect on value of ΔH
A	moved to right	decreased
B	unaffected	increased
C	moved to left	unaffected
D	unaffected	unaffected

25. On the structure shown, four hydrogen atoms have been replaced by letters A, B, C and D.

       (A) H O—(B) H   O
        \ | | | //
         C—C—C———C—C
        // | | \
       O   H H (C)   O—(D)

Which letter corresponds to the hydrogen atom which can ionise most easily in aqueous solution?

26. A fully dissociated acid is progressively diluted by the addition of water.

Which of the following would increase with increasing dilution?

A The pH value
B The electrical conductivity
C The rate of its reaction with chalk
D The volume of alkali which it will neutralise

27. Which of the following is a redox reaction?

A NaOH + HCl → NaCl + H_2O
B Zn + 2HCl → $ZnCl_2$ + H_2
C NiO + 2HCl → $NiCl_2$ + H_2O
D $CuCO_3$ + 2HCl → $CuCl_2$ + H_2O + CO_2

28. If 96 500 C of electricity are passed through separate solutions of copper(II) chloride and nickel(II) chloride, then

A equal masses of copper and nickel will be deposited

B the same number of atoms of each metal will be deposited

C the metals will be plated on the positive electrode

D different numbers of moles of each metal will be deposited.

29. Strontium-90 is a radioisotope.

What is the neutron to proton ratio in an atom of this isotope?

A 2·37

B 1·00

C 0·730

D 1·37

30. Which equation represents a fusion process?

A $^{40}_{19}K + ^{0}_{-1}e \rightarrow ^{40}_{18}Ar$

B $^{2}_{1}H + ^{3}_{1}H \rightarrow ^{4}_{2}He + ^{1}_{0}n$

C $^{235}_{92}U + ^{1}_{0}n \rightarrow ^{90}_{38}Sr + ^{144}_{54}Xe + 2^{1}_{0}n$

D $^{14}_{7}N + ^{1}_{0}n \rightarrow ^{14}_{6}C + ^{1}_{1}p$

[Turn over

Official SQA Past Papers: Higher Chemistry 2002

SECTION A

PART 2

In questions 31 to 35 of this part of the paper, an answer is given by circling the appropriate letter (or letters) in the answer grids provided on Part 2 of the answer sheet.

In some questions, two letters are required for full marks.

If more than the correct number of answers is given, marks may be deducted.

In some cases the number of correct responses is NOT identified in the question.

This part of the paper is worth 10 marks.

SAMPLE QUESTION

A CH_4	B H_2	C CO_2
D CO	E C_2H_6	F N_2

(a) Identify the diatomic **compound(s)**.

A	B	C
(D)	E	F

The one correct answer to part (a) is D. This should be circled.

(b) Identify the **two** substances which burn to produce **both** carbon dioxide **and** water.

(A)	B	C
D	(E)	F

As indicated in this question, there are **two** correct answers to part (b). These are A and E. Both answers are circled.

(c) Identify the substance(s) which can **not** be used as a fuel.

A	B	(C)
D	E	(F)

There are **two** correct answers to part (c). These are C and F. Both answers are circled.

If, after you have recorded your answer, you decide that you have made an error and wish to make a change, you should cancel the original answer and circle the answer you now consider to be correct. Thus, in part (a), if you want to change an answer **D** to an answer **A**, your answer sheet would look like this:

(A)	B	C
D̷	E	F

If you want to change back to an answer which has already been scored out, you should enter a tick (✓) in the box of the answer of your choice, thus:

A̷	B	C
✓D̷	E	F

31. The first twenty elements can be arranged according to bonding and structure.

A aluminium	B boron	C chlorine
D hydrogen	E phosphorus	F silicon

(a) Identify the element which is a discrete molecular solid at room temperature and pressure.

(b) Identify the **two** elements which combine to form the compound with most covalent character.
(You may wish to use page 10 of the data booklet.)

32. Compounds can have different structures and properties.

A NH_4NO_3	B $BaSO_4$	C Na_2CO_3
D SiO_2	E K_2O	F P_2O_5

(a) Identify the compound with a covalent network structure.

(b) Identify the salt which dissolves in water to form an alkaline solution.

[Turn over

33. The symbol for the Avogadro Constant is N_A.
Identify the **true** statement(s).

A	64·2 g of sulphur contains approximately N_A atoms.
B	16·0 g of oxygen contains approximately N_A molecules.
C	6·0 g of water contains approximately N_A atoms.
D	1·0 g of hydrogen contains approximately N_A protons.
E	2·0 litres of 0·50 mol l^{-1} sulphuric acid contains approximately N_A hydrogen ions.
F	1·0 litre of 1·0 mol l^{-1} barium hydroxide solution contains approximately N_A hydroxide ions.

34. Proteins are an important part of a balanced diet.
Identify the **true** statement(s).

A	Proteins are a more concentrated source of energy than carbohydrates.
B	Proteins are made by addition polymerisation.
C	Denaturing of proteins involves changes in the structure of the molecules.
D	Globular proteins are the **major** structural materials of animal tissue.
E	Proteins are compounds of nitrogen, carbon, hydrogen and oxygen.
F	Proteins can be made in animals but **not** in plants.

35. Two flasks contained equal volumes of 0.1 mol l^{-1} hydrochloric acid and 0.1 mol l^{-1} ethanoic acid. Identify the **true** statement(s) about **both** solutions.

A	They give the same colour with Universal indicator.
B	They have a pH less than 7.
C	They conduct electricity equally well.
D	They have equal concentrations of hydrogen ions.
E	They react at the same rate with magnesium.
F	They neutralise the same number of moles of sodium hydroxide.

Candidates are reminded that the answer sheet MUST be returned INSIDE the front cover of this answer book.

[Turn over

SECTION B

1. The three statements below are taken from a note made by a student who is studying trends in the Periodic Table.

   1 **First Ionisation Energy**
      The energy required to remove one mole of electrons from one mole of atoms in the gaseous state.

   2 **Second Ionisation Energy**
      The energy required to remove a second mole of electrons.

   3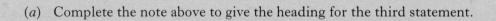
      The measure of the attraction an atom has for the shared electrons in a bond.

   (a) Complete the note above to give the heading for the third statement.

   (b) What is the trend in the first ionisation energy across a period from left to right?

   (c) Why is the second ionisation energy of sodium so much greater than its first ionisation energy?

2. Carbon dating can be used to estimate the age of charcoal found in archaeological sites.

The graph shows how the count rate of a sample of radioactive carbon-14 changes over a period of time.

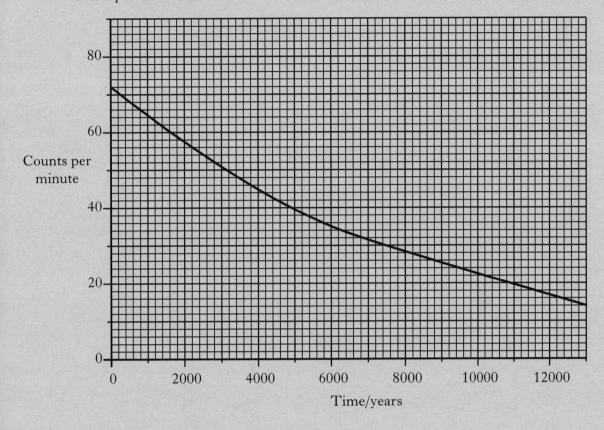

(a) Use the graph to find the half-life of carbon-14.

(b) Carbon-14 decays by beta-emission.
Write the balanced nuclear equation for this decay.

(c) Why can carbon dating **not** be used to estimate the age of fossil fuels?

Marks

3. The structure of a molecule found in olive oil can be represented as shown.

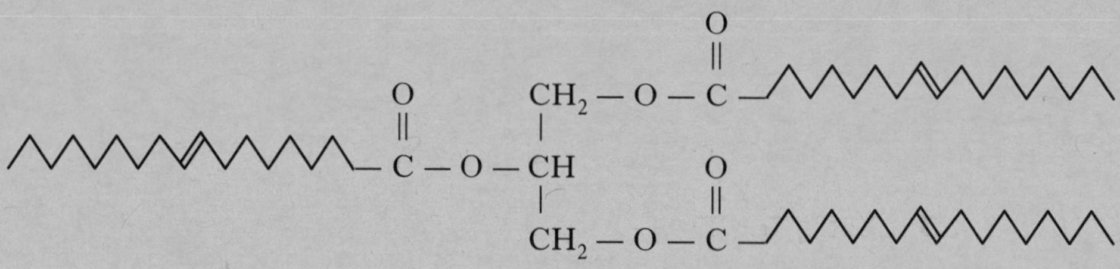

(*a*) Olive oil can be hardened using a nickel catalyst to produce a fat.

(i) What type of catalyst is nickel in this reaction? **1**

heterogenious

(ii) In what way does the structure of a fat molecule differ from that of an oil molecule? **1**

fat molecules have no ~~box~~ double bonds. oil molecules do.

(*b*) Olive oil can be hydrolysed using sodium hydroxide solution to produce sodium salts of fatty acids.

(i) Name the other product of this reaction. **1**

Gylecerol

(ii) Give a commercial use for sodium salts of fatty acids. **1**

soap

(4)

4. Hydrogen sulphide, H_2S, is the unpleasant gas produced when eggs rot.

(a) (i) The gas can be prepared by the reaction of iron(II) sulphide with dilute hydrochloric acid. Iron(II) chloride is the other product of the reaction.

Write a balanced chemical equation for this reaction.

(ii) Iron metal is often present as an impurity in iron(II) sulphide.

Name the other product which would be formed in the reaction with dilute hydrochloric acid if iron metal is present as an impurity.

(b) The enthalpy of combustion of hydrogen sulphide is $-563\,kJ\,mol^{-1}$.

Use this value and the enthalpy of combustion values in the data booklet to calculate the enthalpy change for the reaction:

$$H_2(g) \;+\; S(s) \longrightarrow H_2S(g)$$
$$\text{(rhombic)}$$

Show your working clearly.

5. An ester can be prepared by the following sequence of reactions.

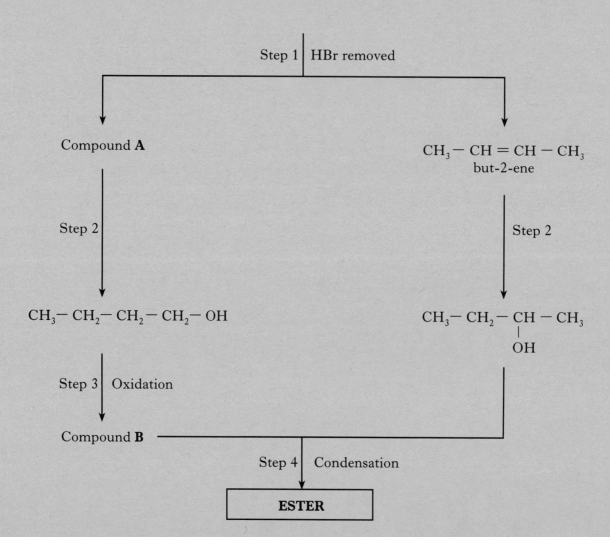

(a) (i) Draw a structural formula for compound **A**.

1

DO NOT WRITE IN THIS MARGIN

Marks

5. (*a*) (continued)

(ii) But-2-ene and compound **A** undergo the same type of reaction in Step 2.
Name this type of reaction. **1**

(iii) Acidified potassium dichromate solution can be used to carry out Step 3.
What colour change would be observed? **1**

(iv) Name compound **B**. **1**

(*b*) (i) What evidence would show that an ester had been formed in Step 4? **1**

(ii) Give **one** use for esters. **1**

(6)

[Turn over

6. A calorimeter, like the one shown, can be used to measure the enthalpy of combustion of ethanol.

The ethanol is ignited and burns completely in the oxygen gas. The heat energy released in the reaction is taken in by the water as the hot product gases are drawn through the coiled copper pipe by the pump.

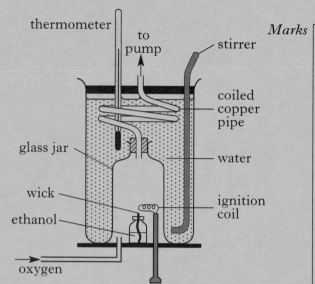

(a) Why is the copper pipe coiled as shown in the diagram?

(b) The value for the enthalpy of combustion of ethanol obtained by the calorimeter method is higher than the value obtained by the typical school laboratory method.

One reason for this is that more heat is lost to the surroundings in the typical school laboratory method.

Give **one** other reason for the value being higher with the calorimeter method.

(c) In one experiment the burning of 0·980 g of ethanol resulted in the temperature of 400 cm³ of water rising from 14·2 °C to 31·6 °C.

Use this information to calculate the enthalpy of combustion of ethanol.

Show your working clearly.

7. A mass spectrometer is an instrument that can be used to gain information about the masses of molecules.

   When hydrogen fluoride is analysed in a mass spectrometer, as well as molecules with a relative molecular mass of 20, some "double molecules" (relative molecular mass 40) and "triple molecules" (relative molecular mass 60) are found to exist. No such molecules are found when the elements, hydrogen and fluorine, are separately analysed.

   (a) Name the weak force of attraction between molecules that is found in both liquid hydrogen and liquid fluorine.

   (b) Why are "double" and "triple" molecules found in hydrogen fluoride but **not** in hydrogen and **not** in fluorine?

8. A student added 0·20 g of silver nitrate, $AgNO_3$, to 25 cm^3 of water. This solution was then added to 20 cm^3 of 0·0010 mol l^{-1} hydrochloric acid as shown in the diagram.

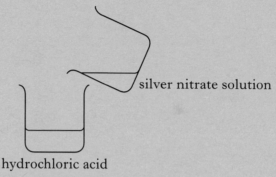

The equation for the reaction which occurs is:

$$AgNO_3(aq) + HCl(aq) \longrightarrow AgCl(s) + HNO_3(aq)$$

(a) (i) Name the type of reaction which takes place.

(ii) Show by calculation which reactant is in excess.
Show your working clearly.

DO NOT WRITE IN THIS MARGIN

Marks

8. (continued)

(*b*) The hydrochloric acid in the experiment can be described as a dilute solution of a strong acid.

(i) What is meant by a strong acid? **1**

(ii) What is the pH of the 0·0010 $mol\,l^{-1}$ hydrochloric acid used in the experiment? **1**

(5)

[Turn over

9. The decomposition of hydrogen peroxide solution into water and oxygen can be catalysed by an enzyme.

$$2H_2O_2(aq) \xrightarrow{enzyme} 2H_2O(\ell) + O_2(g)$$

The rate of reaction can be followed by recording the mass loss over a period of time.

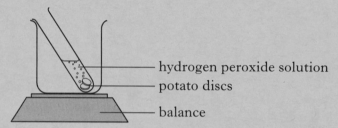

(a) The following graph was obtained from experiments to find the effect of pH on the efficiency of the enzyme.

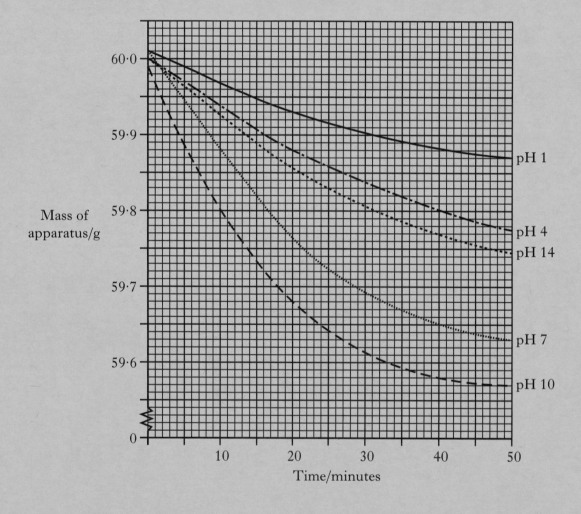

9. (a) (continued)

   (i) Calculate the average rate of reaction over the first 20 minutes, in g min⁻¹, for the experiment at pH 10.

   2

   (ii) From the results shown on the graph, what can be concluded about the efficiency of the enzyme over the pH range used in the experiment?

   1

(b) Give **one** other way of following the rate of this reaction.

1
(4)

[Turn over

10. Consider the following industrial processes.

Contact Process $2SO_2(g) + O_2(g) \rightleftharpoons 2SO_3(g)$ ΔH –ve

Haber Process $N_2(g) + 3H_2(g) \rightleftharpoons 2NH_3(g)$ ΔH –ve

(a) For each process, circle the reactant that can be classified as a raw material.

(b) Explain why increasing the temperature in both processes decreases the equilibrium yield of the products.

(c) Suggest why the Contact Process is carried out at atmospheric pressure but the Haber Process is carried out at 400 atmospheres.

(d) Under certain conditions, 200 kg of hydrogen reacts with excess nitrogen in the Haber Process to produce 650 kg of ammonia.

Calculate the percentage yield of ammonia.

Show your working clearly.

11. Acrylonitrile, CH_2CHCN, is the monomer used in the manufacture of Acrilan.

(a) (i) Draw the full structural formula for acrylonitrile.

(ii) Name the type of polymerisation which occurs in the manufacture of Acrilan.

(b) Acrylonitrile can be reduced in neutral aqueous solution forming $(CH_2CH_2CN)_2$. Hydroxide ions are also produced in the reaction.

Complete and balance the ion-electron equation for the reduction reaction described above.

$$CH_2CHCN \longrightarrow (CH_2CH_2CN)_2 + OH^-$$

[Turn over

12. When sodium hydrogencarbonate is heated to 112 °C it decomposes and the gas, carbon dioxide, is given off:

$$2NaHCO_3(s) \longrightarrow Na_2CO_3(s) + CO_2(g) + H_2O(g)$$

The following apparatus can be used to measure the volume of carbon dioxide produced by the reaction.

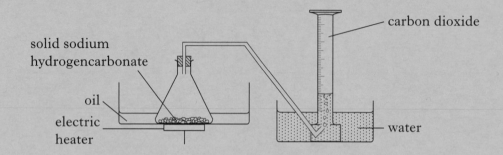

(a) Why is an oil bath used and **not** a water bath?

(b) (i) Calculate the theoretical volume of carbon dioxide produced by the complete decomposition of 1·68 g of sodium hydrogencarbonate.

(Take the molar volume of carbon dioxide to be 23 litre mol^{-1}.)

Show your working clearly.

(ii) Assuming that all of the sodium hydrogencarbonate is decomposed, suggest why the volume of carbon dioxide collected in the measuring cylinder would be less than the theoretical value.

DO NOT WRITE IN THIS MARGIN

Marks

13. Alkenes can react with oxygen to produce unstable compounds called peroxides. These peroxides break down rapidly to form compounds which have the same functional group.

For example, alkene **X** reacts to produce compounds **Y** and **Z**.

(In the following structural formulae R' and R" are used to represent different alkyl groups.)

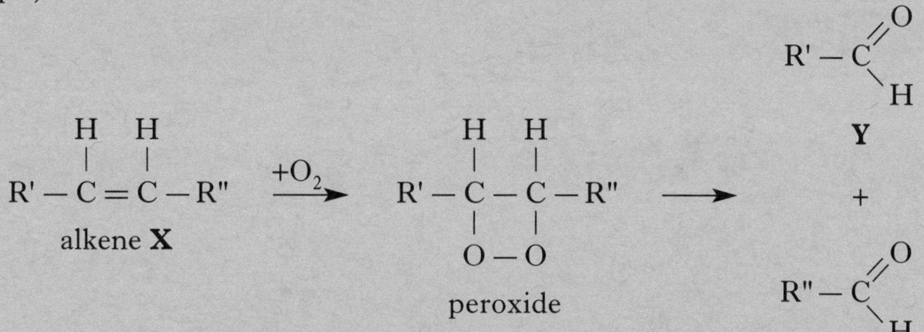

(*a*) To which homologous series do both compounds **Y** and **Z** belong? **1**

(*b*) In one reaction, alkene **X** reacts to produce the two compounds shown below.

Y **Z**

Name alkene **X** in this reaction. **1**

(2)

[Turn over

14. The concentration of a solution of sodium thiosulphate can be found by reaction with iodine.

The iodine is produced by electrolysis of an iodide solution using the apparatus shown.

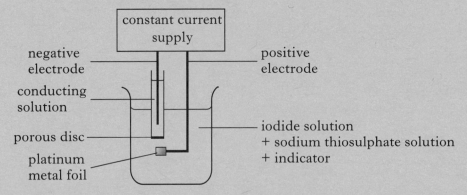

The current is noted and the time when the indicator detects the end-point of the reaction is recorded.

(a) Iodine is produced from the iodide solution according to the following equation:

$$2I^-(aq) \longrightarrow I_2(aq) + 2e^-$$

Calculate the number of moles of iodine generated during the electrolysis given the following results.

Current = 0·010 A
Time = 1 min 37 s

Show your working clearly.

2

14. (continued)

(b) The iodine produced reacts with the thiosulphate ions according to the equation:

$$I_2(aq) + 2S_2O_3^{2-}(aq) \longrightarrow 2I^-(aq) + S_4O_6^{2-}(aq)$$
iodine   thiosulphate ions

At the end-point of the reaction, excess iodine is detected by the indicator.

(i) Name the indicator which could be used to detect the excess iodine present at the end-point.

(ii) In a second experiment it was found that 1.2×10^{-5} mol of iodine reacted with $3.0 \, cm^3$ of the sodium thiosulphate solution.

Use this information to calculate the concentration of the sodium thiosulphate solution, in $mol \, l^{-1}$.

Show your working clearly.

(iii) The production of iodine takes place at the surface of the platinum foil at the tip of the positive electrode.

Suggest what could be done to the solution during the reaction to increase the accuracy of the results.

15. Although they are more expensive, fuel cells have been developed as an alternative to petrol for motor vehicles.

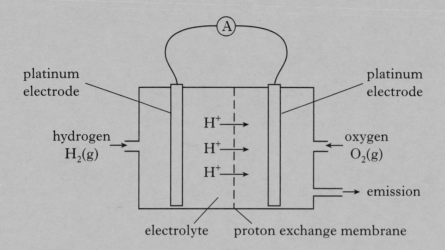

(a) (i) The ion-electron equations for the process occurring at each electrode are:

$$H_2(g) \longrightarrow 2H^+(aq) + 2e^-$$

$$O_2(g) + 4H^+(aq) + 4e^- \longrightarrow 2H_2O(\ell)$$

Combine these two equations to give the overall redox equation.

(ii) On the diagram, show by means of an arrow, the path of electron flow.

(b) Give **one** advantage that fuel cells have over petrol for providing energy.

[END OF QUESTION PAPER]

2003 | Higher

[BLANK]

Official SQA Past Papers: Higher Chemistry 2003

FOR OFFICIAL USE

Total Section B

X012/301

NATIONAL QUALIFICATIONS 2003

FRIDAY, 23 MAY 1.00 PM – 3.30 PM

CHEMISTRY HIGHER

Fill in these boxes and read what is printed below.

Full name of centre

Town

Forename(s)

Surname

Date of birth
Day Month Year

Scottish candidate number

Number of seat

Reference may be made to the Chemistry Higher and Advanced Higher Data Booklet (1999 edition).

SECTION A—Questions 1–40

Instructions for completion of **Section A** are given on page two.

SECTION B

1. All questions should be attempted.

2. The questions may be answered in any order but all answers are to be written in the spaces provided in this answer book, and must be written clearly and legibly in ink.

3. Rough work, if any should be necessary, should be written in this book and then scored through when the fair copy has been written.

4. Additional space for answers and rough work will be found at the end of the book. If further space is required, supplementary sheets may be obtained from the invigilator and should be inserted inside the **front** cover of this book.

5. The size of the space provided for an answer should not be taken as an indication of how much to write. It is not necessary to use all the space.

6. Before leaving the examination room you must give this book to the invigilator. If you do not, you may lose all the marks for this paper.

SCOTTISH QUALIFICATIONS AUTHORITY

SECTION A

1. Check that the answer sheet provided is for Chemistry Higher (Section A).

2. Fill in the details required on the answer sheet.

3. **In questions 1 to 40 of the paper, an answer is given by indicating the choice A, B, C or D by a stroke made in INK in the appropriate place in the answer sheet—see the sample question below.**

4. **For each question there is only ONE correct answer.**

5. Rough working, if required, should be done only on this question paper, or on the rough working sheet provided—**not** on the answer sheet.

6. At the end of the examination the answer sheet for Section A **must** be placed **inside** the front cover of this answer book.

This part of the paper is worth 40 marks.

SAMPLE QUESTION

To show that the ink in a ball-pen consists of a mixture of dyes, the method of separation would be

    A fractional distillation

    B chromatography

    C fractional crystallisation

    D filtration.

The correct answer is B—chromatography. A **heavy** vertical line should be drawn joining the two dots in the appropriate box in the column headed **B** as shown **in the example on the answer sheet**.

If, after you have recorded your answer, you decide that you have made an error and wish to make a change, you should cancel the original answer and put a vertical stroke in the box you now consider to be correct. Thus, if you want to change an answer **D** to an answer **B**, your answer sheet would look like this:

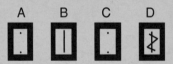

If you want to change back to an answer which has already been scored out, you should **enter a tick (✓)** to the RIGHT of the box of your choice, thus:

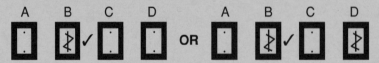

1. Which of the following substances is a non-conductor when solid, but becomes a good conductor on melting?

   A Argon

   B Potassium

   C Potassium fluoride

   D Tetrachloromethane

2. Which of the following covalent gases does **not** react with water forming ions?

   A HCl

   B SO_2

   C NH_3

   D CH_4

3. An iron nail is covered with water.

   Which of the following procedures would **not** increase the rate at which the iron nail corrodes?

   A Adding some sodium sulphate to the water

   B Adding some glucose to the water

   C Attaching a copper wire to the nail

   D Passing carbon dioxide through the water

4. Naturally occurring nitrogen consists of two isotopes ^{14}N and ^{15}N.

   How many different types of nitrogen molecules will occur in the air?

   A 1

   B 2

   C 3

   D 4

5. A mixture of sodium chloride and sodium sulphate is known to contain 0·6 mol of chloride ions and 0·2 mol of sulphate ions.

   How many moles of sodium ions are present?

   A 0·4

   B 0·5

   C 0·8

   D 1·0

6. When copper carbonate reacts with excess acid, carbon dioxide is produced. The curves shown were obtained under different conditions.

   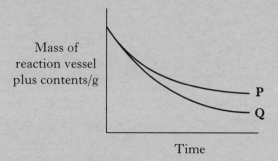

   The change from **P** to **Q** could be brought about by

   A increasing the concentration of the acid

   B increasing the mass of copper carbonate

   C decreasing the particle size of the copper carbonate

   D adding a catalyst.

7. When 3·6 g of butanal (relative formula mass = 72) was burned, 134 kJ of energy was released.

   From this result, what is the enthalpy of combustion, in $kJ\,mol^{-1}$?

   A −6·7

   B +6·7

   C −2680

   D +2680

8. Which of the following chlorides is likely to have the **most** ionic character?

   A LiCl

   B CsCl

   C $BeCl_2$

   D $CaCl_2$

9. Which equation represents the first ionisation energy of a diatomic element, X_2?

   A $\tfrac{1}{2}X_2(s) \rightarrow X^+(g)$

   B $\tfrac{1}{2}X_2(g) \rightarrow X^-(g)$

   C $X(g) \rightarrow X^+(g)$

   D $X(s) \rightarrow X^-(g)$

[Turn over

10. Which of the following elements exists as discrete molecules?

    A Boron

    B Carbon (diamond)

    C Silicon

    D Sulphur

11. Which of the following chlorides is most likely to be soluble in tetrachloromethane, CCl_4?

    A Barium chloride

    B Caesium chloride

    C Calcium chloride

    D Phosphorus chloride

12. In which of the following liquids does hydrogen bonding occur?

    A Ethanoic acid

    B Ethyl ethanoate

    C Hexane

    D Hex-1-ene

13. A compound boils at $-33\,°C$. It also dissolves in water to give an alkaline solution.

    Which type of bonding is present within the compound?

    A Metallic

    B Covalent (polar)

    C Ionic

    D Covalent (non-polar)

14. The number of moles of ions in 1 mol of copper(II) phosphate is

    A 1

    B 2

    C 4

    D 5.

15. The Avogadro Constant is the same as the number of

    A atoms in 24 g of carbon

    B molecules in 16 g of oxygen

    C molecules in 2 g of hydrogen

    D ions in 1 litre of sodium chloride solution, concentration $1\ mol\,l^{-1}$.

16. What volume of oxygen, in litres, is required for the complete combustion of 1 litre of butane gas?

    (All volumes are measured under the same conditions of temperature and pressure.)

    A 1

    B 4

    C 6·5

    D 13

17. Which of the following processes can be used industrially to produce aromatic hydrocarbons?

    A Reforming of naphtha

    B Catalytic cracking of propane

    C Reforming of coal

    D Catalytic cracking of heavy oil fractions

18. Which line in the table refers to a hydrocarbon that is **not** a member of the same homologous series as the others?

	Relative formula mass
A	44
B	72
C	84
D	100

19. Which of the following compounds does **not** have isomeric structures?

    A C_2HCl_3

    B $C_2H_4Cl_2$

    C Propene

    D Propan-1-ol

20. Which of the following structural formulae represents a tertiary alcohol?

A $\quad CH_3 - \underset{\underset{CH_3}{|}}{\overset{\overset{CH_3}{|}}{C}} - CH_2 - OH$

B $\quad CH_3 - \underset{\underset{OH}{|}}{\overset{\overset{CH_3}{|}}{C}} - CH_2 - CH_3$

C $\quad CH_3 - CH_2 - CH_2 - \underset{\underset{OH}{|}}{\overset{\overset{H}{|}}{C}} - CH_3$

D $\quad CH_3 - CH_2 - \underset{\underset{OH}{|}}{\overset{\overset{H}{|}}{C}} - CH_2 - CH_3$

21. Oxidation of 4-methylpentan-2-ol to the corresponding ketone results in the alcohol

A losing 2 g per mole
B gaining 2 g per mole
C gaining 16 g per mole
D not changing in mass.

22. What type of reaction takes place when propene is formed from propanol?

A Condensation
B Hydrolysis
C Dehydration
D Hydration

23. The extensive use of which type of compound is thought to contribute significantly to the depletion of the ozone layer?

A Oxides of carbon
B Hydrocarbons
C Oxides of sulphur
D Chlorofluorocarbons

24. Polyesters can exist as fibres and cured resins.

Which line in the table describes correctly the structure of their molecules?

	Polyester fibre	Cured polyester resin
A	cross-linked	cross-linked
B	linear	linear
C	cross-linked	linear
D	linear	cross-linked

25. Ammonia solution may be used to distinguish $Fe^{2+}(aq)$ from $Fe^{3+}(aq)$ as follows:

$Fe^{2+}(aq)$ gives a green precipitate of $Fe(OH)_2$;

$Fe^{3+}(aq)$ gives a brown precipitate of $Fe(OH)_3$.

Which of the following is most likely to give similar results if used instead of ammonia?

A An amine
B An alcohol
C An aldehyde
D A carboxylic acid

26. In the formation of "hardened" fats from vegetable oils, the hydrogen

A causes cross-linking between chains
B causes hydrolysis to occur
C increases the carbon chain length
D reduces the number of carbon-carbon double bonds.

[Turn over

27. Olestra is a calorie free fat made by reacting fatty acids with sucrose. The structure of a sucrose molecule can be represented as shown.

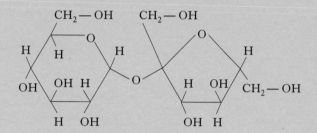

How many fatty acid molecules can react with one molecule of sucrose?

A 3
B 5
C 8
D 11

28. Proteins can be denatured under acid conditions.

During this denaturing, the protein molecule

A changes shape
B is dehydrated
C is neutralised
D is polymerised.

29. Which of the following substances is a raw material for the chemical industry?

A Benzene
B Methane
C Aluminium
D Iron

30. What is the relationship between enthalpies p, q, r and s?

$S(s) + H_2(g) \rightarrow H_2S(g)$ $\Delta H = p$

$H_2(g) + \tfrac{1}{2}O_2(g) \rightarrow H_2O(\ell)$ $\Delta H = q$

$S(s) + O_2(g) \rightarrow SO_2(g)$ $\Delta H = r$

$H_2S(g) + 1\tfrac{1}{2}O_2(g) \rightarrow H_2O(\ell) + SO_2(g)$ $\Delta H = s$

A $p = q + r - s$
B $p = s - q - r$
C $p = q - r - s$
D $p = s + r - q$

31. A catalyst is added to a reaction at equilibrium.

Which of the following does **not** apply?

A The rate of the forward reaction increases.
B The rate of the reverse reaction increases.
C The position of equilibrium remains unchanged.
D The position of equilibrium shifts to the right.

32. $ICl(\ell) + Cl_2(g) \rightleftharpoons ICl_3(s)$ $\Delta H = -106 \, kJ\,mol^{-1}$

Which line in the table identifies correctly the changes that will cause the greatest increase in the proportion of solid in the above equilibrium?

	Temperature	Pressure
A	decrease	decrease
B	decrease	increase
C	increase	decrease
D	increase	increase

33. Which of the following is the best description of a $0.1 \, mol\,l^{-1}$ solution of ethanoic acid?

A Dilute solution of a weak acid
B Dilute solution of a strong acid
C Concentrated solution of a weak acid
D Concentrated solution of a strong acid

34. The concentration of $OH^-(aq)$ ions in a solution is 1×10^{-2} mol l^{-1}.

What is the concentration of $H^+(aq)$ ions, in mol l^{-1}?

A 1×10^{-2}

B 1×10^{-5}

C 1×10^{-9}

D 1×10^{-12}

35. When a certain aqueous solution is diluted, its conductivity decreases but its pH remains constant.

The solution could be

A ethanoic acid

B sodium chloride

C sodium hydroxide

D nitric acid.

36. The ion-electron equations for a redox reaction are:

$$2I^-(aq) \rightarrow I_2(aq) + 2e^-$$
$$MnO_4^-(aq) + 8H^+(aq) + 5e^- \rightarrow Mn^{2+}(aq) + 4H_2O(\ell)$$

How many moles of iodide ions are oxidised by one mole of permanganate ions?

A 0·2

B 0·4

C 2

D 5

37. In which of the following reactions is hydrogen gas acting as an oxidising agent?

A $H_2 + C_2H_4 \rightarrow C_2H_6$

B $H_2 + Cl_2 \rightarrow 2HCl$

C $H_2 + 2Na \rightarrow 2NaH$

D $H_2 + CuO \rightarrow H_2O + Cu$

38. When 10 g of lead pellets containing radioactive lead are placed in a solution containing 10 g of lead nitrate, radioactivity soon appears in the solution.

Compared to the pellets the solution will show

A different intensity of radiation and different half-life

B the same intensity of radiation but different half-life

C different intensity of radiation but the same half-life

D the same intensity of radiation and the same half-life.

39. Which line in the table describes correctly the result of an atom losing a beta-particle?

$^{0}_{-1}\beta$ $^{0}_{-1}e$

	Atomic number	Mass number
A	increased	no change
B	decreased	no change
C	no change	increased
D	no change	decreased

40. $^{2}_{1}H + ^{3}_{1}H \rightarrow ^{4}_{2}He + ^{1}_{0}n$

The above process represents

A nuclear fission

B nuclear fusion

C proton capture

D neutron capture.

Candidates are reminded that the answer sheet MUST be returned INSIDE the front cover of this answer book.

[Turn over

SECTION B

1. Unleaded petrol uses hydrocarbons with a high degree of molecular branching in order to improve the efficiency of burning.

   The structure of one such hydrocarbon is shown.

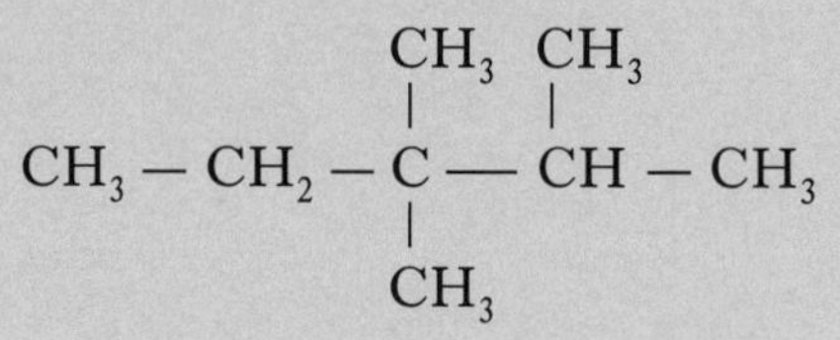

   $$CH_3-CH_2-\underset{\underset{CH_3}{|}}{\overset{\overset{CH_3}{|}}{C}}-\overset{\overset{CH_3}{|}}{CH}-CH_3$$

   (a) Give the systematic name for this hydrocarbon.

   2,3,3-trimethylpentane

   (b) Name **one** other **type** of hydrocarbon that is used in petrol for the same reason.

   cycloalkane

2. Concorde aircraft were grounded after an incident in which the fuel tank in one of the aeroplanes was punctured by a piece of metal. As a result the fuel tanks are now coated with the polymer, Kevlar.

(a) What property of Kevlar makes it suitable for this use?

It is strong but also light

(b) The repeating unit in Kevlar is shown.

What name is given to the outlined group in this repeating unit?

the amide link / peptide link

3. Polonium-210 is a radioisotope that decays by alpha-emission.
The half-life of polonium-210 is 140 days.

(a) Draw a graph to show how the mass of 200 g of the radioisotope would change with time.
(Additional graph paper, if required, can be found on page 32.)

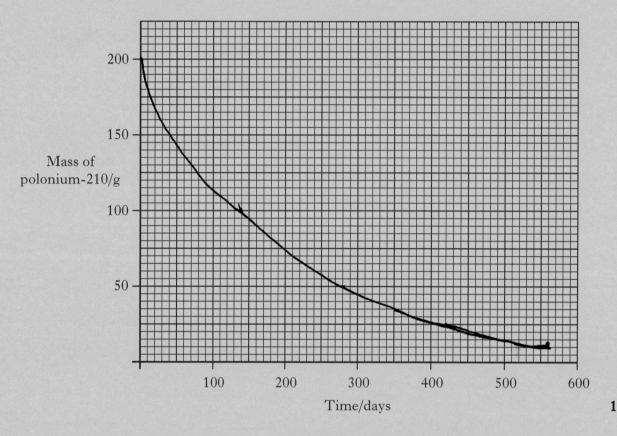

(b) Write a balanced nuclear equation for the alpha-decay of polonium-210.

$$^{210}_{84}Po \rightarrow {}^{206}_{82}Pb + {}^{4}_{2}\alpha$$

(c) Calculate the number of atoms in 105 g of polonium-210.

4. The Thermite Process involves the reaction between aluminium and iron(III) oxide to produce iron and aluminium oxide.

This highly exothermic reaction, which generates so much heat that the temperature of the mixture rises to around 3000 °C, is used for repairing cracked railway lines as shown in the diagram below.

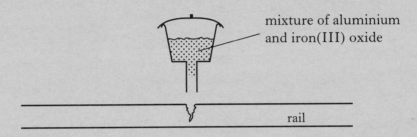

(a) Suggest why this process is suitable for repairing cracked railway lines.

(b) The enthalpy changes for the formation of one mole of aluminium oxide and one mole of iron(III) oxide are shown below.

$$2Al(s) + 1\tfrac{1}{2}O_2(g) \longrightarrow Al_2O_3(s) \quad \Delta H = -1676 \text{ kJ mol}^{-1}$$
$$2Fe(s) + 1\tfrac{1}{2}O_2(g) \longrightarrow Fe_2O_3(s) \quad \Delta H = -825 \text{ kJ mol}^{-1}$$

Use the above information to calculate the enthalpy change for the reaction:

$$2Al(s) + Fe_2O_3(s) \longrightarrow Al_2O_3(s) + 2Fe(s)$$

[Turn over

5. Urea-methanal is a polymer which can be made using coal as a feedstock.

The flow diagram shows the steps involved in the production of the polymer.

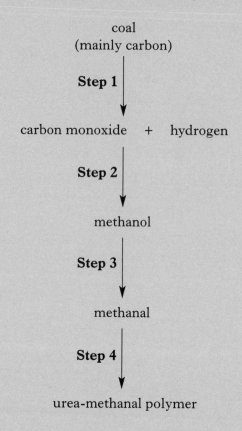

(a) Name the mixture of gases produced in **Step 1**.

(b) Name the type of reaction taking place in **Step 3**.

DO NOT WRITE IN THIS MARGIN

Marks

5. (continued)

(*c*) (i) In **Step 4** methanal reacts with urea, H_2NCONH_2.

Draw the full structural formula for urea. **1**

(ii) The urea-methanal polymer does **not** soften on heating.

What name is given to this type of plastic? **1**

(4)

[Turn over

6. Ethyl pentanoate is an ester. It can be prepared in the lab as shown below.

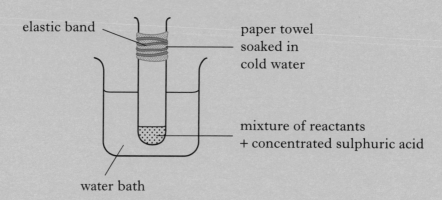

(a) (i) Why is a water bath used for heating?

(ii) What is the purpose of the wet paper towel?

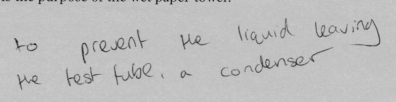

to prevent the liquid leaving the test tube, a condenser

(b) Draw a structural formula for ethyl pentanoate.

DO NOT WRITE IN THIS MARGIN

Marks

6. (continued)

(*c*) Starting with a mass of 3·6 g of ethanol, and a slight excess of pentanoic acid, a student achieved a 70% yield of ethyl pentanoate (mass of one mole = 130 g).

Calculate the mass of ester obtained.

Show your working clearly.

2

(5)

[Turn over

7. A student added 50 cm³ of 4·0 mol l⁻¹ hydrochloric acid to 4·0 g of magnesium ribbon.

(a) The balanced equation for the reaction is:

$$Mg(s) + 2HCl(aq) \longrightarrow MgCl_2(aq) + H_2(g)$$

Show by calculation which reactant was in excess.
Show your working clearly.

(b) The hydrogen produced in the reaction can be contaminated with small quantities of hydrogen chloride vapour.

This vapour is very soluble in water.

Complete the diagram to show how the hydrogen chloride can be removed before the hydrogen is collected.

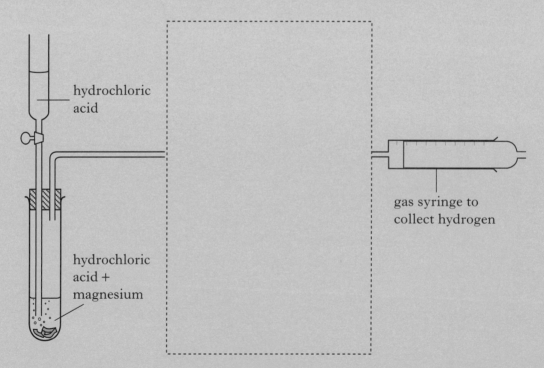

DO NOT WRITE IN THIS MARGIN

Marks

7. (continued)

(*c*) The experiment was repeated under the same conditions using ethanoic acid instead of hydrochloric acid.

Circle the correct words in the table to show the results for ethanoic acid.

	Ethanoic acid
Rate of reaction	faster/same/slower
Volume of gas produced	more/same/less

1

(4)

[Turn over

8. Although aldehydes and ketones have different structures, they both contain the carbonyl functional group.

(a) (i) In what way is the structure of an aldehyde different from that of a ketone?

(ii) As a result of the difference in structure, aldehydes react with Fehling's (or Benedict's) solution and Tollens' reagent but ketones do not.

What colour change would be observed when propanal is heated with Fehling's (or Benedict's) solution?

(iii) In the reaction of propanal with Tollens' reagent, silver ions are reduced to form silver metal.

Complete the following ion-electron equation for the oxidation.

$$C_3H_6O \longrightarrow C_2H_5COOH$$

(iv) Name the compound with the formula C_2H_5COOH.

8. (continued)

(b) As a result of both containing the carbonyl group, aldehydes and ketones react in a similar way with hydrogen cyanide.

The equation for the reaction of propanal and hydrogen cyanide is shown.

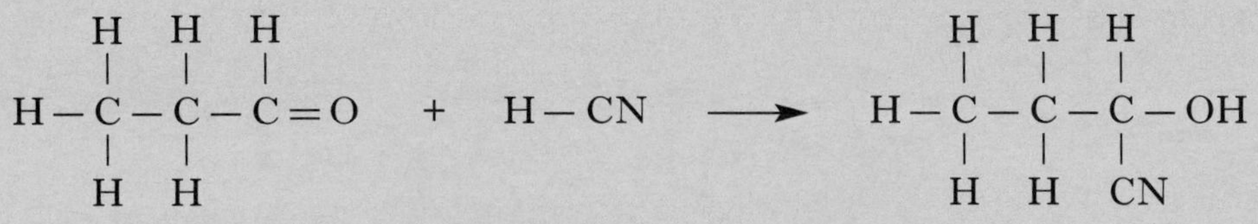

```
    H   H   H                              H   H   H
    |   |   |                              |   |   |
H — C — C — C = O   +   H — CN   ⟶   H — C — C — C — OH
    |   |                                  |   |   |
    H   H                                  H   H   CN
```

(i) Suggest a name for this type of reaction.

addition

1

(ii) Draw a structure for the product of the reaction between propanone and hydrogen cyanide.

1
(6)

[Turn over

9. A student was asked to write a plan of the procedure for an investigation. The entry made in her laboratory note book is shown.

> **Aim**
>
> **To find the effect of concentration on the rate of the reaction between hydrogen peroxide and an acidified solution of iodide ions.**
>
> $$H_2O_2(aq) + 2H^+(aq) + 2I^-(aq) \rightarrow 2H_2O(\ell) + I_2(aq)$$
>
> **Procedure**
>
> 1. Using a 100 cm³ measuring cylinder, measure out 10 cm³ of sulphuric acid, 10 cm³ of sodium thiosulphate solution, 1 cm³ of starch solution and 25 cm³ of potassium iodide solution into a dry 100 cm³ glass beaker and place the beaker on the bench.
>
> 2. Measure out 5 cm³ of hydrogen peroxide solution and start the timer.
>
> 3. Add the hydrogen peroxide solution to the beaker. When the blue/black colour just appears, stop the timer and record the time (in seconds).
>
> 4. Repeat this procedure four times but using different concentrations of potassium iodide solution. This is achieved by adding 5 cm³, 10 cm³, 15 cm³ and 20 cm³ of water to the 25 cm³ of potassium iodide solution before adding it to the glass beaker.

(a) Why is instruction 4 **not** the best way of altering the concentration of the potassium iodide solution?

(b) State **two** other ways of improving the student's plan of this investigation procedure.

10. If both potassium iodide solution, KI(aq), and liquid chloroform, $CHCl_3(\ell)$, are added to a test-tube with some iodine, the iodine dissolves in both. Two layers are formed as shown in the diagram.

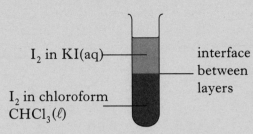

An equilibrium is set up:

$$I_2 \text{ in KI(aq)} \rightleftharpoons I_2 \text{ in } CHCl_3(\ell)$$

The iodine is always distributed between the two layers in the same ratio:

$$\frac{\text{concentration of } I_2 \text{ in } CHCl_3(\ell)}{\text{concentration of } I_2 \text{ in KI(aq)}} = \frac{3}{1}$$

(a) What is meant by the term **equilibrium**?

(b) When more potassium iodide solution is added to the top layer the equilibrium is disturbed.

What happens to restore the equilibrium?

(c) 0·4 g of I_2 is dissolved in 10 cm³ of KI(aq) and 10 cm³ of $CHCl_3(\ell)$.

Calculate the concentration of iodine, **in g l⁻¹**, contained in $CHCl_3(\ell)$.

11. Hydrogen peroxide, H_2O_2, decomposes very slowly to produce water and oxygen.

(a) The activation energy (E_A) for the reaction is **75 kJ mol^{-1}** and the enthalpy change (ΔH) is **–26 kJ mol^{-1}**.

Use this information to complete the potential energy diagram for the reaction.
(Additional graph paper, if required, can be found on page 32.)

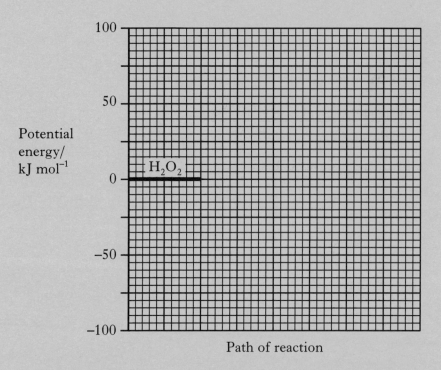

Path of reaction

(b) Powdered manganese dioxide catalyses the decomposition of hydrogen peroxide solution.

(i) What name is given to this type of catalyst?

(ii) Add a dotted line to the above diagram to show the path of the reaction when the catalyst is used.

11. (continued)

(c) The balanced equation for the reaction is:

$$2H_2O_2(aq) \longrightarrow 2H_2O(\ell) + O_2(g)$$

(i) The following graph is obtained for the volume of oxygen released over time.

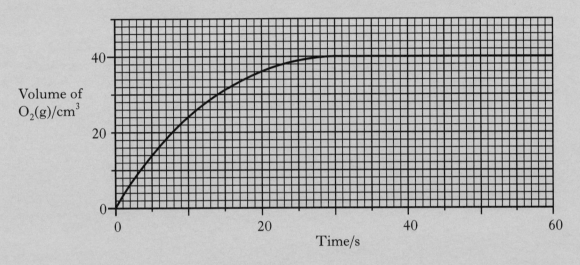

Calculate the average rate of reaction between 10 and 20 s.

(ii) Using information from the above graph, calculate the mass of hydrogen peroxide used in the reaction, assuming all the hydrogen peroxide decomposed.

(Take the molar volume of oxygen to be 24 litres mol^{-1}.)

Show your working clearly.

12. Although propane and ethanol have similar molecular masses, the alkane is a gas at room temperature while the alcohol is a liquid. This difference in boiling points is due to the different strengths of the intermolecular forces in the two compounds.

Explain why propane is a gas at room temperature while ethanol is a liquid.

In your answer you should name the intermolecular forces involved in each compound and explain how they arise.

(4)

DO NOT WRITE IN THIS MARGIN

Marks

13. Potassium cyanide, KCN, can be made by the reaction of an acid with an alkali.

A solution of the salt has a pH of 8.

(*a*) What is the concentration of $H^{+}(aq)$, in $mol\,l^{-1}$, in the solution? **1**

(*b*) What can be concluded about the strengths of the acid and the alkali used in the reaction? **1**

(*c*) Write the formula for the acid used in the reaction. **1**

(3)

[Turn over

14. The following flow diagram outlines the manufacture of sodium carbonate by the Solvay Process.

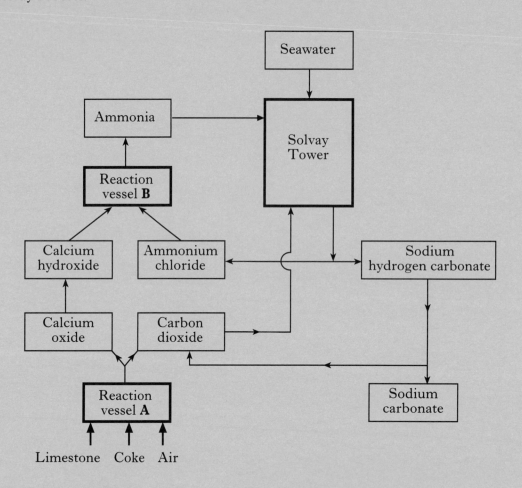

(a) Name the reactants in the reaction taking place in the Solvay Tower.

(b) In reaction vessel **A**, carbon dioxide is produced by the following two reactions.

$CaCO_3(s) \rightarrow CaO(s) + CO_2(g)$ $\Delta H = $ _____

$C(s) + O_2(g) \rightarrow CO_2(g)$ $\Delta H = $ _____

For each reaction, add a sign after the ΔH to show whether the reaction is endothermic or exothermic.

DO NOT WRITE IN THIS MARGIN

Marks

14. (continued)

(*c*) As well as ammonia, a salt and water are produced in reaction vessel **B**.

Write a balanced equation for the production of ammonia in this reaction vessel. **1**

(*d*) The seawater used in the Solvay Process can contain contaminant magnesium ions. These can be removed by the addition of sodium carbonate solution.

Why is sodium carbonate solution suitable for removing contaminant magnesium ions? **1**

(*e*) Using the information in the flow diagram, give **two different** features of the Solvay Process that make it economical. **2**

(6)

[Turn over

15. Electrophoresis, widely used in medicine and forensic testing, involves the movement of ions in an electric field. The technique can be used to separate and identify amino acids produced by the breakdown of proteins.

(a) Name the type of reaction that takes place during the breakdown of proteins.

Marks 1

(b) (i) The amino acid, glycine, has the following structural formula:

$$\begin{array}{c} NH_2 \\ | \\ H-C-H \\ | \\ COOH \end{array}$$

Like all amino acids, glycine exists as ions in solution and the charge on the ions depends on the pH of the solution. In solutions with **low** pH, glycine exists as a **positively** charged ion:

$$\begin{array}{c} NH_3^+ \\ | \\ H-C-H \\ | \\ COOH \end{array}$$

In solutions with a **high** pH, glycine exists as a **negatively** charged ion. Draw the structure of this negatively charged ion.

1

15. (b) (continued)

(ii) The table below shows the structures and molecular masses of three amino acids, **A**, **B** and **C**.

Amino acid	Structure	Molecular mass
A	H – C(NH$_2$)(COOH) – CH$_2$ – C$_6$H$_5$	165·0
B	H – C(NH$_2$)(COOH) – CH$_2$CH$_2$CH$_2$CH$_2$ – NH$_2$	146·0
C	H – C(NH$_2$)(COOH) – CH$_2$CH$_2$ – COOH	147·0

A mixture of amino acids, **A**, **B** and **C**, was applied to the centre of a strip of filter paper which had been soaked in a solution of pH 2. All three amino acids exist as ions in this acidic solution. A high voltage was then applied across the filter paper.

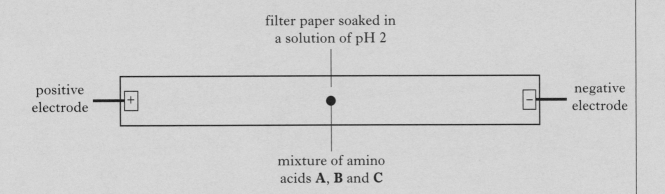

The amino acid ions separate according to their **charge** and **molecular mass**.

On the diagram above, indicate the approximate positions of **A**, **B** and **C** once electrophoresis has separated the ions.

2

(4)

[Turn over

16. A student electrolysed dilute sulphuric acid using the apparatus shown in order to estimate the volume of one mole of hydrogen gas.

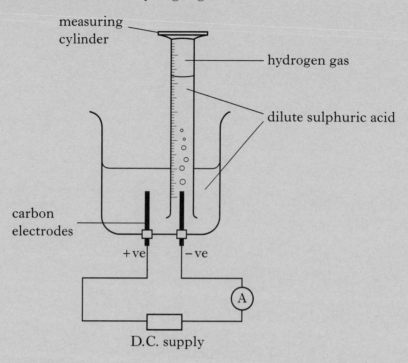

(a) The measurements recorded by the student were:

Current = 0·5 A
Time = 14 minutes
Volume of hydrogen collected = 52 cm³

Calculate the molar volume of hydrogen gas.
Show your working clearly.

(3)

(b) What change could be made to the apparatus to reduce a possible significant source of error?

(1)

[END OF QUESTION PAPER]

ADDITIONAL SPACE FOR ANSWERS

ADDITIONAL GRAPH PAPER FOR QUESTION 3(a)

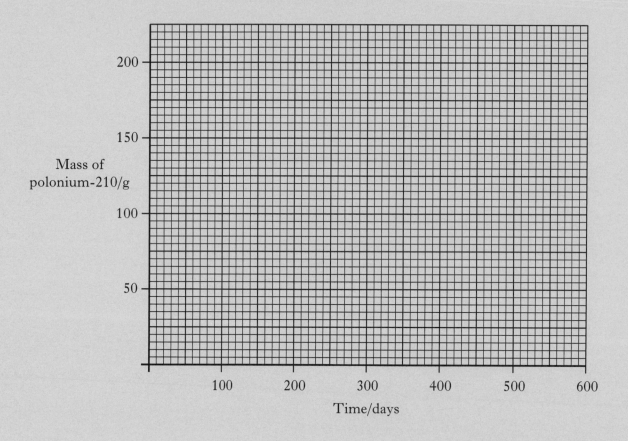

ADDITIONAL GRAPH PAPER FOR QUESTION 11(a)

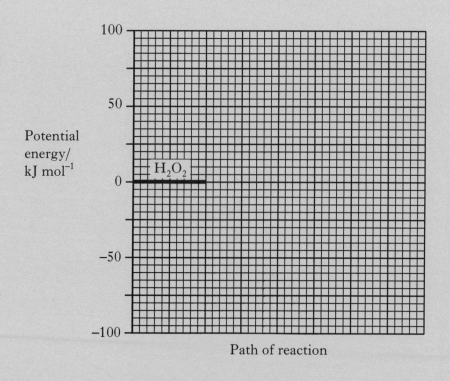

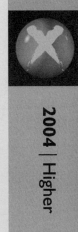

2004 | Higher

[BLANK]

X012/301

NATIONAL QUALIFICATIONS 2004

WEDNESDAY, 2 JUNE 9.00 AM – 11.30 AM

CHEMISTRY HIGHER

Fill in these boxes and read what is printed below.

Full name of centre

Town

Forename(s)

Surname

Date of birth
Day Month Year   Scottish candidate number   Number of seat

Reference may be made to the Chemistry Higher and Advanced Higher Data Booklet (1999 edition).

SECTION A—Questions 1–40

Instructions for completion of **Section A** are given on page two.

SECTION B

1. All questions should be attempted.

2. The questions may be answered in any order but all answers are to be written in the spaces provided in this answer book, and must be written clearly and legibly in ink.

3. Rough work, if any should be necessary, should be written in this book and then scored through when the fair copy has been written.

4. Additional space for answers and rough work will be found at the end of the book. If further space is required, supplementary sheets may be obtained from the invigilator and should be inserted inside the **front** cover of this book.

5. The size of the space provided for an answer should not be taken as an indication of how much to write. It is not necessary to use all the space.

6. Before leaving the examination room you must give this book to the invigilator. If you do not, you may lose all the marks for this paper.

SECTION A

1. Check that the answer sheet provided is for Chemistry Higher (Section A).

2. Fill in the details required on the answer sheet.

3. In questions 1 to 40 of the paper, an answer is given by indicating the choice A, B, C or D by a stroke made in INK in the appropriate place in the answer sheet—see the sample question below.

4. **For each question there is only ONE correct answer.**

5. Rough working, if required, should be done only on this question paper, or on the rough working sheet provided—**not** on the answer sheet.

6. At the end of the examination the answer sheet for Section A **must** be placed **inside** the front cover of this answer book.

This part of the paper is worth 40 marks.

SAMPLE QUESTION

To show that the ink in a ball-pen consists of a mixture of dyes, the method of separation would be

    A fractional distillation

    B chromatography

    C fractional crystallisation

    D filtration.

The correct answer is B—chromatography. A **heavy** vertical line should be drawn joining the two dots in the appropriate box in the column headed **B** as shown **in the example on the answer sheet**.

If, after you have recorded your answer, you decide that you have made an error and wish to make a change, you should cancel the original answer and put a vertical stroke in the box you now consider to be correct. Thus, if you want to change an answer **D** to an answer **B**, your answer sheet would look like this:

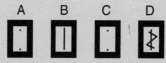

If you want to change back to an answer which has already been scored out, you should **enter a tick (✓)** to the RIGHT of the box of your choice, thus:

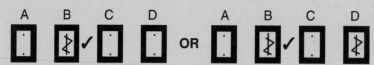

1. Which of the following solids has a low melting point and a high electrical conductivity?

   A Iodine
   B Potassium
   C Silicon oxide
   D Potassium fluoride

2. Two experiments are set up to study the corrosion of an iron nail.

   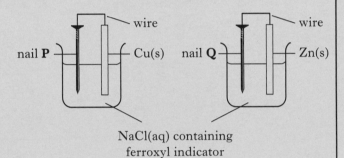

   After a short time, a blue colour will have appeared at

   A both **P** and **Q**
   B neither **P** nor **Q**
   C **P** but not at **Q**
   D **Q** but not at **P**.

3. In which of the following compounds do **both** ions have the same electron arrangement as argon?

   A Calcium sulphide
   B Magnesium oxide
   C Sodium sulphide
   D Calcium bromide

4. What volume of sodium hydroxide solution, concentration $0.4\ mol\,l^{-1}$, is needed to neutralise $50\ cm^3$ of sulphuric acid, concentration $0.1\ mol\,l^{-1}$?

   A $25\ cm^3$
   B $50\ cm^3$
   C $100\ cm^3$
   D $200\ cm^3$

5. Like atoms, molecules can lose electrons to form positive ions.

   1. $[^1H_2{}^{16}O]^+$ 2. $[^1H_2{}^{17}O]^+$ 3. $[^1H_2{}^{18}O]^+$
   4. $[^2H_2{}^{16}O]^+$ 5. $[^2H_2{}^{17}O]^+$ 6. $[^2H_2{}^{18}O]^+$

   Which of the following pairs has ions of the same mass?

   A 1 and 4
   B 2 and 5
   C 3 and 6
   D 3 and 4

6. Which of the following graphs could represent the change in the rate of a reaction between magnesium ribbon and hydrochloric acid?

   A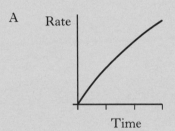

   B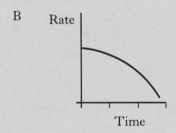

   C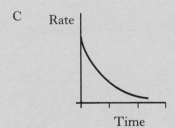

   D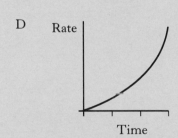

[Turn over

7. 1 mol of hydrogen gas and 1 mol of iodine vapour were mixed and allowed to react. After t seconds, 0·8 mol of hydrogen remained.

   The number of moles of hydrogen iodide formed at t seconds was

   A 0·2

   B 0·4

   C 0·8

   D 1·6.

8.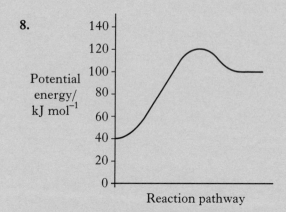

   Reaction pathway

   Which of the following sets of data applies to the reaction represented by the above energy diagram?

	Enthalpy change	Activation energy/ kJ mol^{-1}
A	Exothermic	60
B	Exothermic	80
C	Endothermic	60
D	Endothermic	80

9. Which of the following elements has the greatest electronegativity?

   A Caesium

   B Oxygen

   C Fluorine

   D Iodine

10. As the relative atomic mass in the halogens increases

    A the boiling point increases

    B the density decreases

    C the first ionisation energy increases

    D the atomic size decreases.

11. Which of the following elements would require the most energy to convert one mole of gaseous atoms into gaseous ions each carrying two positive charges?

    (You may wish to use the data booklet.)

    A Scandium

    B Titanium

    C Vanadium

    D Chromium

12. Which of the following compounds has polar molecules?

    A CH_4

    B CO_2

    C NH_3

    D CCl_4

13. $2NO(g) + O_2(g) \rightarrow 2NO_2(g)$

    How many litres of nitrogen dioxide gas could theoretically be obtained in the reaction of 1 litre of nitrogen monoxide gas with 2 litres of oxygen gas?

    (All volumes are measured under the same conditions of temperature and pressure.)

    A 1

    B 2

    C 3

    D 4

14. Which of the following gases has the same volume as 128·2 g of sulphur dioxide gas?

    (All volumes are measured under the same conditions of temperature and pressure.)

    A 2·0 g of hydrogen

    B 8·0 g of helium

    C 32·0 g of oxygen

    D 80·8 g of neon

15. 5 g of copper is added to excess silver(I) nitrate solution. After some time, the solid present is filtered off from the copper(II) nitrate solution, washed with water, dried, and weighed.

The final mass of the solid will be

A less than 5 g

B 5 g

C 10 g

D more than 10 g.

16. Which of the following equations represents a reaction which takes place during reforming?

A $C_6H_{14} \rightarrow C_6H_6 + 4H_2$

B $C_4H_8 + H_2 \rightarrow C_4H_{10}$

C $C_2H_5OH \rightarrow C_2H_4 + H_2O$

D $C_8H_{18} \rightarrow C_4H_{10} + C_4H_8$

17. Which of the following is a ketone?

A C₆H₅–CHO

B C₆H₅–CO–CH₃

C C₆H₅–COOH

D C₆H₅–COO–CH₃

18. Which of the following is an isomer of 2,2-dimethylpentan-1-ol?

A $CH_3CH_2CH_2CH(CH_3)CH_2OH$

B $(CH_3)_3CCH(CH_3)CH_2OH$

C $CH_3CH_2CH_2CH_2CH_2CH_2CH_2CH_2OH$

D $(CH_3)_2CHC(CH_3)_2CH_2CH_2OH$

19. Ethene is used in the manufacture of addition polymers.

What type of reaction is used to produce ethene from ethane?

A Addition

B Cracking

C Hydrogenation

D Oxidation

20. The compound $CH_3CH_2COO^-Na^+$ is formed by reaction between sodium hydroxide and

A propanoic acid

B propan-1-ol

C propene

D propanal.

21. Which of the following is **not** a correct statement about methanol?

A It is a primary alkanol.

B It can be oxidised to methanal.

C It can be made from synthesis gas.

D It can be dehydrated to an alkene.

22. Ammonia is manufactured from hydrogen and nitrogen by the Haber Process

$$3H_2 + N_2 \rightleftharpoons 2NH_3$$

If 80 kg of ammonia is produced from 60 kg of hydrogen, what is the percentage yield?

A $\dfrac{80}{340} \times 100$

B $\dfrac{80}{170} \times 100$

C $\dfrac{30}{80} \times 100$

D $\dfrac{60}{80} \times 100$

23. What mixture of gases is known as synthesis gas?

A Methane and oxygen

B Carbon monoxide and oxygen

C Carbon dioxide and hydrogen

D Carbon monoxide and hydrogen

24. Part of a polymer chain is shown below.

$$-O-\overset{O}{\underset{\|}{C}}-(CH_2)_4-\overset{O}{\underset{\|}{C}}-O-(CH_2)_6-O-\overset{O}{\underset{\|}{C}}-(CH_2)_4-\overset{O}{\underset{\|}{C}}-O-(CH_2)_6-O-$$

Which of the following compounds, when added to the reactants during polymerisation, would stop the polymer chain from getting too long?

A $\quad HO-\overset{O}{\underset{\|}{C}}-(CH_2)_4-\overset{O}{\underset{\|}{C}}-OH$

B $\quad HO-(CH_2)_6-OH$

C $\quad HO-(CH_2)_5-\overset{O}{\underset{\|}{C}}-OH$

D $\quad CH_3-OH$

---

25. Which of the following polymers is used in making bullet-proof vests?

A Kevlar

B Biopol

C Poly(ethenol)

D Poly(ethyne)

26. Which of the following is a structural formula for glycerol?

A $\quad \begin{array}{c} CH_2OH \\ | \\ CHOH \\ | \\ CH_2OH \end{array}$

B $\quad \begin{array}{c} CH_2OH \\ | \\ CH_2 \\ | \\ CH_2OH \end{array}$

C $\quad \begin{array}{c} CH_2OH \\ | \\ CH_2OH \end{array}$

D $\quad \begin{array}{c} CH_2OH \\ | \\ CHOH \\ | \\ CH_2COOH \end{array}$

27. Fats have higher melting points than oils because comparing fats and oils

A fats have more hydrogen bonds

B fat molecules are more saturated

C fat molecules are more loosely packed

D fats have more cross-links between molecules.

28. The monomer units used to construct enzyme molecules are

A alcohols

B esters

C amino acids

D fatty acids.

29. Which of the following compounds is a raw material in the chemical industry?

A Ethene

B Ammonia

C Sulphuric acid

D Sodium chloride

30. Consider the reaction pathway shown.

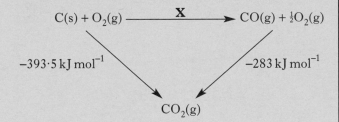

According to Hess's Law, what is the enthalpy change for reaction **X**?

A $+110 \cdot 5 \text{ kJ mol}^{-1}$

B $-110 \cdot 5 \text{ kJ mol}^{-1}$

C $-676 \cdot 5 \text{ kJ mol}^{-1}$

D $+676 \cdot 5 \text{ kJ mol}^{-1}$

31. Which line in the table applies correctly to the use of a catalyst in a chemical reaction?

	Position of equilibrium	Effect on value of ΔH
A	Moved to right	Decreased
B	Unaffected	Increased
C	Moved to left	Unaffected
D	Unaffected	Unaffected

32. Some solid ammonium chloride is added to a dilute solution of ammonia.

Which of the following ions will decrease in concentration as a result?

A Ammonium

B Hydrogen

C Hydroxide

D Chloride

33. The pH of a solution of hydrochloric acid was found to be 2·5.

The concentration of the $H^+(aq)$ ions in the acid must be

A greater than $0 \cdot 1 \text{ mol l}^{-1}$

B between $0 \cdot 1$ and $0 \cdot 01 \text{ mol l}^{-1}$

C between $0 \cdot 01$ and $0 \cdot 001 \text{ mol l}^{-1}$

D less than $0 \cdot 001 \text{ mol l}^{-1}$.

34. Which of the following is the best description of a $0 \cdot 1 \text{ mol l}^{-1}$ solution of sulphuric acid?

A Dilute solution of a strong acid

B Dilute solution of a weak acid

C Concentrated solution of a strong acid

D Concentrated solution of a weak acid

35. Excess marble chips (calcium carbonate) were added to 100 cm^3 of 1 mol l^{-1} hydrochloric acid. The experiment was repeated using the same mass of the marble chips and 100 cm^3 of 1 mol l^{-1} ethanoic acid.

Which of the following would have been the same for both experiments?

A The time taken for the reaction to be completed

B The rate at which the first 10 cm^3 of gas was evolved

C The mass of marble chips left over when the reaction had stopped

D The average rate of the reaction

36. Which line in the table is correct for $0 \cdot 1 \text{ mol l}^{-1}$ sodium hydroxide compared with $0 \cdot 1 \text{ mol l}^{-1}$ aqueous ammonia?

	pH	Conductivity
A	higher	lower
B	higher	higher
C	lower	higher
D	lower	lower

37. During a redox process in acid solution, iodate ions, $IO_3^-(aq)$, are converted into iodine, $I_2(aq)$.

$12H^+ + 2IO_3^-(aq) \rightarrow I_2(aq) + 6H_2O$

The numbers of $H^+(aq)$ and $H_2O(l)$ required to balance the ion-electron equation for the formation of 1 mol of $I_2(aq)$ are, respectively

A 3 and 6

B 6 and 3

C 6 and 12

D 12 and 6.

38. Ammonia reacts with magnesium as shown.

$$3Mg(s) + 2NH_3(g) \rightarrow (Mg^{2+})_3(N^{3-})_2(s) + 3H_2(g)$$

In this reaction, ammonia is acting as

A an acid

B a base

C an oxidising agent

D a reducing agent.

39. Induced nuclear reactions can be described in a shortened form

$$T(x, y)P$$

where the participants are the target nucleus (T), the bombarding particle (x), the ejected particle (y) and the product nucleus (P).

Which of the following nuclear reactions would **not** give the product nucleus indicated?

A $^{14}_{7}N \quad (\alpha, p) \quad ^{17}_{8}O$

B $^{236}_{93}Np \quad (p, \alpha) \quad ^{238}_{92}U$

C $^{10}_{5}B \quad (\alpha, n) \quad ^{13}_{7}N$

D $^{242}_{96}Cf \quad (n, \alpha) \quad ^{239}_{94}Pu$

40. Which of the following equations represents a nuclear fission process?

A $\quad ^{40}_{19}K + ^{0}_{-1}e \rightarrow ^{40}_{18}Ar$

B $\quad ^{2}_{1}H + ^{3}_{1}H \rightarrow ^{4}_{2}He + ^{1}_{0}n$

C $\quad ^{235}_{92}U + ^{1}_{0}n \rightarrow ^{90}_{38}Sr + ^{144}_{54}Xe + 2^{1}_{0}n$

D $\quad ^{14}_{7}N + ^{1}_{0}n \rightarrow ^{14}_{6}C + ^{1}_{1}p$

Candidates are reminded that the answer sheet MUST be returned INSIDE the front cover of this answer book.

[Turn over for SECTION B on *Page ten*

SECTION B

1. (a) Complete the table below by adding the name of an element from **elements 1 to 20** of the Periodic Table for each of the types of bonding and structure described.

Bonding and structure at room temperature and pressure	Name of element
metallic solid	sodium
monatomic gas	
covalent network solid	
discrete covalent molecular gas	
discrete covalent molecular solid	

(b) Why do metallic solids such as sodium conduct electricity?

2. Phosphorus-32 is a radioisotope that decays by beta-emission.

(a) Write the nuclear equation for the decay of phosphorus-32.

(b) (i) An 8 g sample of phosphorus-32 was freshly prepared.
Calculate the number of phosphorus atoms contained in the 8 g sample.

(ii) The half-life of phosphorus-32 is 14·3 days.
Calculate the time it would take for the mass of phosphorus-32 in the 8 g sample to fall to 1 g.

3. Sphalerite is an impure zinc sulphide ore, containing traces of other metal compounds.

The flow diagram for the extraction of zinc from this ore is shown below.

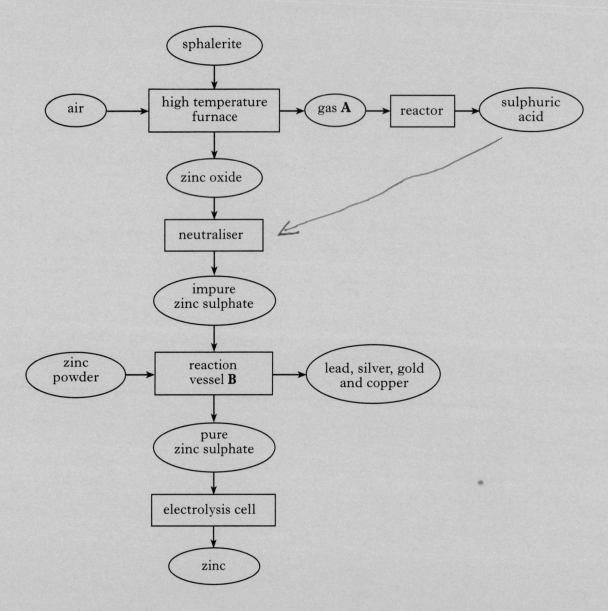

(a) Name gas **A**.

sulphur dioxide

(b) Name the type of reaction taking place in reaction vessel **B**.

Redox

DO NOT WRITE IN THIS MARGIN

Marks

3. (continued)

(*c*) It is economical to make use of the sulphuric acid produced.

Add an arrow to the flow diagram to show how the sulphuric acid could be used in this extraction. **1**

(*d*) The ion-electron equation for the production of zinc in the electrolysis cell is

$$Zn^{2+} + 2e^{-} \longrightarrow Zn$$

If a current of 2000 A is used in the cell, calculate the mass of zinc, in kg, produced in 24 hours.

Show your working clearly. **2**

(5)

[Turn over

4. Glucose is produced in plants by photosynthesis.

(a) Plants convert glucose into a condensation polymer for storing energy.

Name this condensation polymer.

(b) One way of representing the structure of glucose in aqueous solution is shown below.

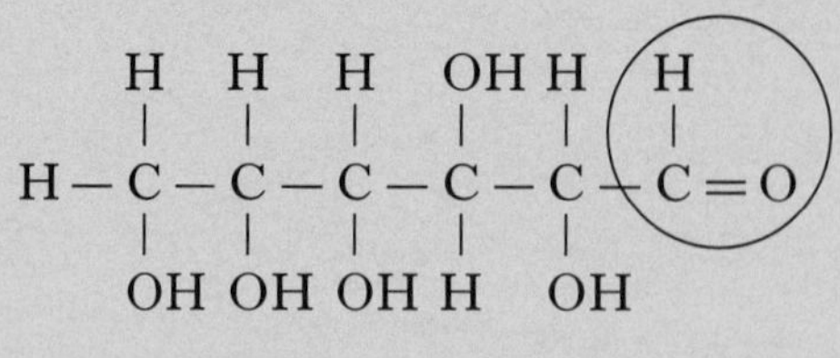

$$H-\underset{OH}{\overset{H}{C}}-\underset{OH}{\overset{H}{C}}-\underset{OH}{\overset{H}{C}}-\underset{H}{\overset{OH}{C}}-\underset{OH}{\overset{H}{C}}-\overset{H}{C}=O$$

In this structure the aldehyde group is circled.

(i) What would be seen when glucose is oxidised using Tollens' reagent?

(ii) Complete the structure below to show the product formed when glucose is oxidised.

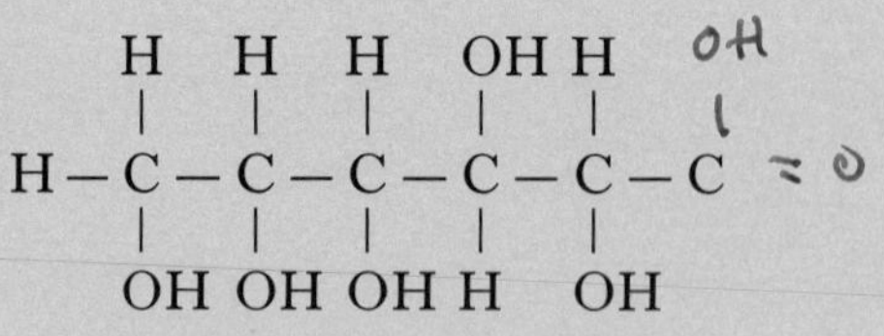

$$H-\underset{OH}{\overset{H}{C}}-\underset{OH}{\overset{H}{C}}-\underset{OH}{\overset{H}{C}}-\underset{H}{\overset{OH}{C}}-\underset{OH}{\overset{H}{C}}-C$$

(c) Under anaerobic conditions, carbohydrates, like glucose, can be used to produce biogas. The main constituent of biogas is methane which is a useful fuel.

State **one** advantage of using biogas as a fuel rather than natural gas.

5. If the conditions are kept constant, reversible reactions will attain a state of equilibrium.

(a) Circle the correct words in the table to show what is true for reactions at equilibrium.

Rate of forward reaction compared to rate of reverse reaction	faster / same / slower
Concentrations of reactants compared to concentrations of products	usually different / always the same

(b) The following equilibrium involves two compounds of phosphorus.

$$PCl_3(g) + 3NH_3(g) \rightleftharpoons P(NH_2)_3(g) + 3HCl(g)$$

(i) An increase in temperature moves the above equilibrium to the left.
What does this indicate about the enthalpy change for the forward reaction?

(ii) What effect, if any, will an increase in pressure have on the above equilibrium?

6. The effect of temperature on reaction rate can be studied using the reaction between oxalic acid and acidified potassium permanganate solutions.

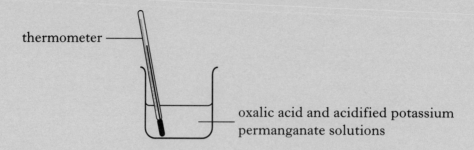

(a) What colour change would indicate that the reaction was complete?

(b) A student's results are shown on the graph below.

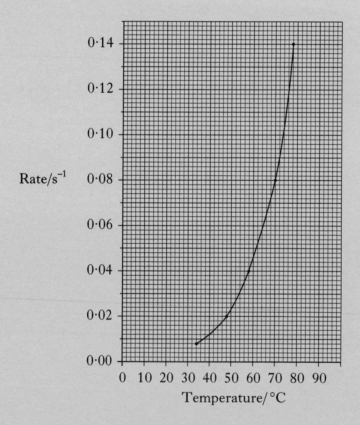

(i) Use the graph to calculate the reaction time, in s, at 40 °C.

6. **(b) (continued)**

(ii) Why is it difficult to obtain an accurate reaction time when the reaction is carried out at room temperature?

1

(c) The diagram below shows the energy distribution of molecules in a gas at a particular temperature.

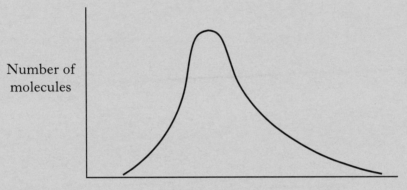

Kinetic energy of the molecules

Draw a second curve on the diagram to show the energy distribution of the molecules in the gas at a higher temperature.

Label the diagram to indicate why an increase in temperature has such a significant effect on reaction rate.

1
(4)

[Turn over

7. Enzymes are specific biological catalysts. For example, trypsin, an enzyme produced in the pancreas, will catalyse the hydrolysis of only certain peptide links in a protein.

(a) Draw the structure of a peptide link.

(b) Trypsin has an optimum temperature of 37 °C.

Draw a curve to show how the enzyme activity varies with temperature.

(Additional graph paper, if required, can be found on page 32.)

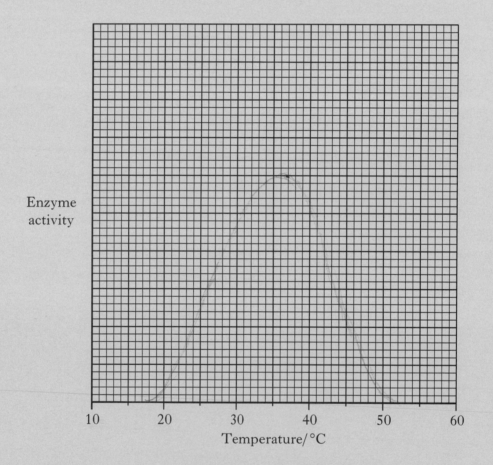

(c) Trypsin loses its activity if placed in a solution of very high pH.
What happens to the enzyme to cause this loss of activity?

DO NOT WRITE IN THIS MARGIN

Marks

8. An experiment using dilute hydrochloric acid and sodium hydroxide solution was carried out to determine the enthalpy of neutralisation.

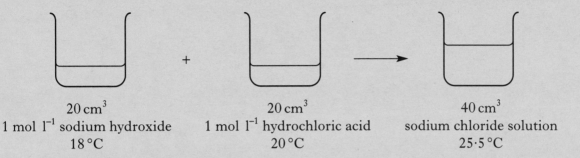

20 cm³ 1 mol l^{-1} sodium hydroxide 18 °C + 20 cm³ 1 mol l^{-1} hydrochloric acid 20 °C → 40 cm³ sodium chloride solution 25·5 °C

(*a*) Using the information in the diagram, calculate the enthalpy of neutralisation, in kJ mol^{-1}.

Show your working clearly. **2**

(*b*) Calculate the concentration of hydroxide ions in the 1 mol l^{-1} hydrochloric acid used in the experiment. **1**

(3)

[Turn over

9. (a) Aluminium and phosphorus are close to one another in the Periodic Table but the P^{3-} ion is much larger than the Al^{3+} ion.

Give the reason for this difference.

(b) The P^{3-} ion and the Ca^{2+} ion have the same electron arrangement but the Ca^{2+} ion is smaller than the P^{3-} ion.

Give the reason for this difference.

10. The structures of two antiseptics are shown. Both are aromatic.

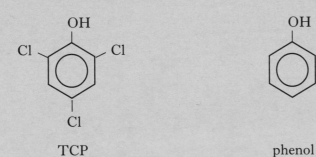

(a) (i) What gives the aromatic ring its stability?

The delocalised inside ring

(ii) Write the molecular formula for TCP.

(iii) The systematic name for TCP is 2,4,6-trichlorophenol.
The systematic name for Dettol, another antiseptic, is 4-chloro-3,5-dimethylphenol.
Draw a structural formula for Dettol.

(b) The feedstocks for the production of antiseptics are made by reforming the naphtha fraction of crude oil.
Give another use for reformed naphtha.

11. A student used the slow reaction between magnesium and water to determine the molar volume of hydrogen gas.

$$Mg(s) + 2H_2O(\ell) \longrightarrow Mg(OH)_2(aq) + H_2(g)$$

The following items were used in the experiment.

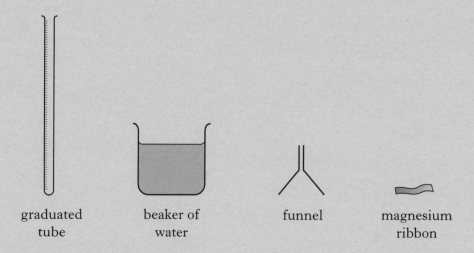

graduated tube   beaker of water   funnel   magnesium ribbon

(a) Draw a diagram to show how the student would have arranged the above items at the start of the experiment.

DO NOT WRITE IN THIS MARGIN

Marks

11. (continued)

(*b*) What measurements would the student take and how would they be used to calculate the **molar** volume of hydrogen gas? **2**

(3)

[Turn over

12. A student added vitamin C solution to iodine solution.

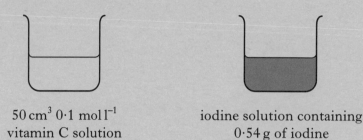

50 cm³ 0·1 mol l⁻¹ vitamin C solution

iodine solution containing 0·54 g of iodine

The equation for the reaction of vitamin C ($C_6H_8O_6$) with iodine solution is shown below.

$$C_6H_8O_6(aq) + I_2(aq) \longrightarrow C_6H_6O_6(aq) + 2H^+(aq) + 2I^-(aq)$$
$$\text{(brown)} \qquad\qquad\qquad\qquad\qquad\qquad \text{(colourless)}$$

(a) By calculating which reactant was in excess, state whether the iodine solution would have been decolourised.

Show your working clearly.

(b) Write the ion-electron equation for the oxidation of vitamin C.

13. Compound **X** is a secondary alcohol.

```
     H   H   H   H
     |   |   |   |
 H – C – C – C – C – H
     |   |   |   |
     H   H   OH  H
```
compound **X**

(a) Name compound **X**.

(b) Draw a structural formula for the tertiary alcohol that is an isomer of compound **X**.

(c) When passed over heated aluminium oxide, compound **X** is dehydrated, producing isomeric compounds, **Y** and **Z**.

Both compounds **Y** and **Z** react with hydrogen bromide, HBr. Compound **Y** reacts to produce two products while compound **Z** reacts to produce only one product.

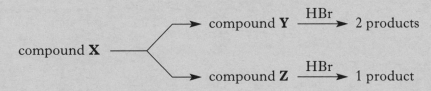

Name compound **Z**.

14. (a) Ethanol and propanoic acid can react to form an ester.

(i) Draw a structural formula for this ester.

(ii) Draw a labelled diagram of the assembled apparatus that could be used to prepare this ester in the laboratory.

(iii) Due to hydrogen bonding, ethanol and propanoic acid are soluble in water whereas the ester produced is insoluble.

In each of the boxes below, draw a molecule of water and use a dotted line to show where a hydrogen bond could exist between the organic molecule and the water molecule.

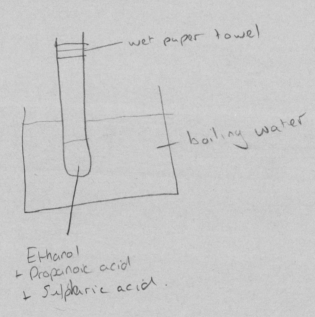

14. (continued)

(b) Pyrolysis (thermal decomposition) of esters can produce two compounds, an alkene and an alkanoic acid, according to the following equation.

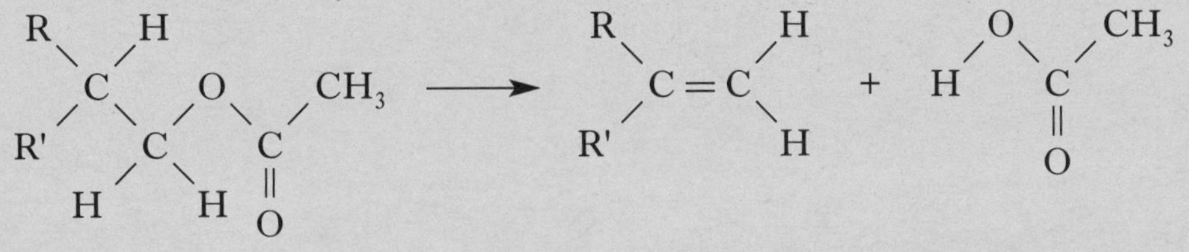

[Diagram showing pyrolysis equation with R, R', H groups]

(R and R' represent alkyl groups)

Draw a structural formula for the ester that would produce 2-methylbut-1-ene and methanoic acid on pyrolysis.

1

(5)

DO NOT WRITE IN THIS MARGIN

Marks

15. Vinegar is a dilute solution of ethanoic acid.

(*a*) Hess's Law can be used to calculate the enthalpy change for the formation of ethanoic acid from its elements.

$$2C(s) + 2H_2(g) + O_2(g) \rightarrow CH_3COOH(\ell)$$
(graphite)

Calculate the enthalpy change for the above reaction, in kJ mol^{-1}, using information from the data booklet and the following data.

$$CH_3COOH(\ell) + 2O_2(g) \rightarrow 2CO_2(g) + 2H_2O(\ell) \quad \Delta H = -876 \text{ kJ mol}^{-1}$$

Show your working clearly.

2

(*b*) Ethanoic acid can be used to prepare the salt, sodium ethanoate, CH_3COONa.

Explain why sodium ethanoate solution has a pH greater than 7.

In your answer you should mention the **two** equilibria involved.

3

(5)

16. Potassium permanganate is a very useful chemical in the laboratory.

(a) Solid potassium permanganate can be heated to release oxygen gas. This reaction can be represented by the equation shown below.

$$KMnO_4(s) \longrightarrow K_2O(s) + MnO_2(s) + O_2(g)$$

Balance the above equation.

(b) An acidified potassium permanganate solution can be used to determine the concentration of a solution of iron(II) sulphate by a titration method.

(i) Apart from taking accurate measurements, suggest **two** points of good practice that a student should follow to ensure that an accurate end-point is achieved in a titration.

(ii) In a titration, a student found that an average of $16 \cdot 7 \text{ cm}^3$ of iron(II) sulphate solution was needed to react completely with $25 \cdot 0 \text{ cm}^3$ of $0 \cdot 20 \text{ mol l}^{-1}$ potassium permanganate solution.

The equation for the reaction is:

$$5Fe^{2+}(aq) + MnO_4^-(aq) + 8H^+(aq) \rightarrow 5Fe^{3+}(aq) + Mn^{2+}(aq) + 4H_2O(\ell)$$

Calculate the concentration of the iron(II) sulphate solution, in mol l^{-1}.

Show your working clearly.

17. A proton NMR spectrum can be used to help identify the structure of an organic compound.

The three key principles used in identifying a group containing hydrogen atoms in a molecule are as follows:

- The position of the line(s) on the x-axis of the spectrum is a measure of the "chemical shift" of the hydrogen atoms in the particular group.

  Some common "chemical shift" values are given in the table below.

Group containing hydrogen atoms	Chemical shift
$-CH_3$	1·0
$-C\equiv CH$	2·7
$-CH_2Cl$	3·7
$-CHO$	9·0

- The number of lines for the hydrogen atoms in the group is $n + 1$ where n is the number of hydrogen atoms on the carbon atom next to the group.
- The maximum height of the line(s) for the hydrogen atoms in the group is relative to the number of hydrogen atoms in the group.

The spectrum for ethanal is shown below.

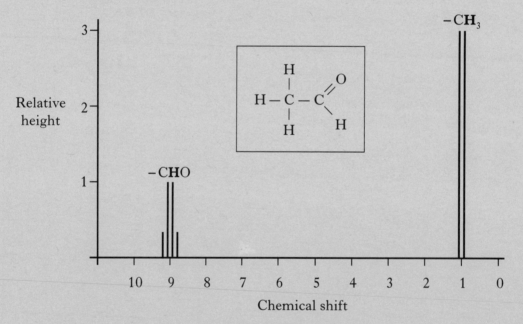

(a) The chemical shift values shown in the table are based on the range of values shown in the data booklet for proton NMR spectra.

Use the data booklet to find the range in the chemical shift values for hydrogen atoms in the following environment:

$$\begin{array}{c}\diagdown\\C=C\\\diagup\end{array}\begin{array}{c}\diagup\\\diagdown\\H\end{array}$$

1

17. (continued)

(b) A carbon compound has the following spectrum.

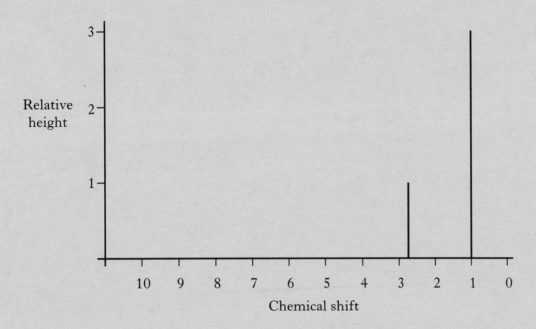

Name this compound.

1

(c) Draw the spectrum that would be obtained for chloroethane.

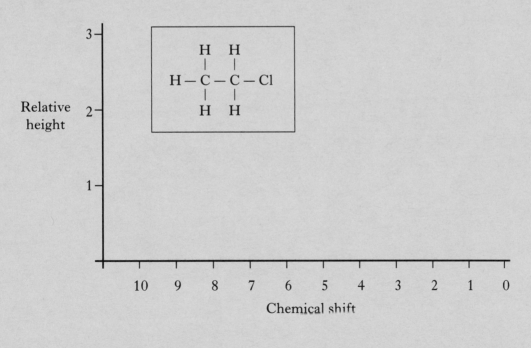

1

(3)

[END OF QUESTION PAPER]

ADDITIONAL SPACE FOR ANSWERS

ADDITIONAL GRAPH PAPER FOR QUESTION 7(*b*)

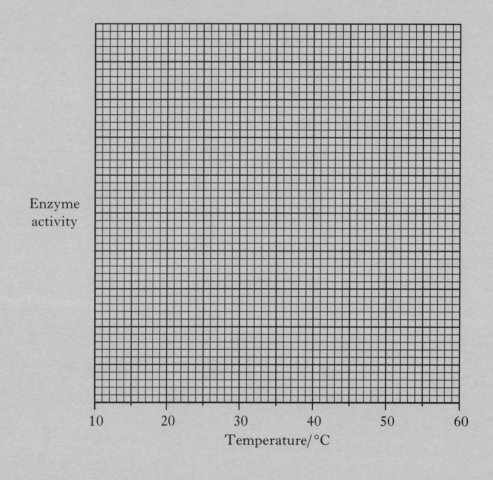

2005 | Higher

[BLANK PAGE]

Official SQA Past Papers: Higher Chemistry 2005

FOR OFFICIAL USE

Total Section B

X012/301

NATIONAL QUALIFICATIONS 2005

TUESDAY, 31 MAY 9.00 AM – 11.30 AM

CHEMISTRY HIGHER

Fill in these boxes and read what is printed below.

Full name of centre

Town

Forename(s)

Surname

Date of birth
Day Month Year   Scottish candidate number   Number of seat

Reference may be made to the Chemistry Higher and Advanced Higher Data Booklet (1999 edition).

SECTION A—Questions 1–40 (40 marks)

Instructions for completion of **Section A** are given on page two.

SECTION B (60 marks)

1 All questions should be attempted.

2 The questions may be answered in any order but all answers are to be written in the spaces provided in this answer book, and must be written clearly and legibly in ink.

3 Rough work, if any should be necessary, should be written in this book and then scored through when the fair copy has been written. If further space is required, a supplementary sheet for rough work may be obtained from the invigilator.

4 Additional space for answers will be found at the end of the book. If further space is required, supplementary sheets may be obtained from the invigilator and should be inserted inside the **front** cover of this book.

5 The size of the space provided for an answer should not be taken as an indication of how much to write. It is not necessary to use all the space.

6 Before leaving the examination room you must give this book to the invigilator. If you do not, you may lose all the marks for this paper.

SCOTTISH QUALIFICATIONS AUTHORITY

SAB X012/301 6/12070

SECTION A

Read carefully

1. Check that the answer sheet provided is for **Chemistry Higher (Section A)**.
2. Check that the answer sheet you have been given has **your name, date of birth, SCN** (Scottish Candidate Number) and **Centre Name** printed on it.
   Do not change any of these details.
3. If any of this information is wrong, tell the Invigilator immediately.
4. If this information is correct, **print** your name and seat number in the boxes provided.
5. Use **black** or **blue ink** for your answers. **Do not use red ink**.
6. The answer to each question is **either** A, B, C or D. Decide what your answer is, then put a horizontal line in the space provided (see sample question below).
7. There is **only one correct** answer to each question.
8. Any rough working should be done on the question paper or the rough working sheet, **not** on your answer sheet.
9. At the end of the exam, put the **answer sheet for Section A inside the front cover of your answer book**.

Sample Question

To show that the ink in a ball-pen consists of a mixture of dyes, the method of separation would be

  A fractional distillation

  B chromatography

  C fractional crystallisation

  D filtration.

The correct answer is **B**—chromatography. The answer **B** has been clearly marked with a horizontal line (see below).

Changing an answer

If you decide to change your answer, cancel your first answer by putting a cross through it (see below) and fill in the answer you want. The answer below has been changed to **B**.

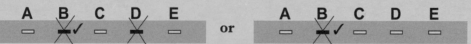

If you then decide to change back to an answer you have already scored out, put a tick (✓) to the **right** of the answer you want, as shown below:

1. Isotopes of an element have

   A the same mass number
   B the same number of neutrons
   C equal numbers of protons and neutrons
   D different numbers of neutrons.

2. Which of the following pairs of solutions would react to produce a precipitate?

   A Barium nitrate and sodium chloride
   B Barium hydroxide and potassium nitrate
   C Copper(II) sulphate and sodium carbonate
   D Copper(II) chloride and potassium sulphate

3. Dilute hydrochloric acid, concentration $2\ mol\ l^{-1}$, is added to a mixture of copper metal and copper(II) carbonate.

   Which of the following happens?

   A The only gas produced is carbon dioxide.
   B The only gas produced is hydrogen.
   C A mixture of carbon dioxide and hydrogen is produced.
   D There is no production of gas.

4. How many moles of magnesium will react with $20\ cm^3$ of $2\ mol\ l^{-1}$ hydrochloric acid?

   $Mg(s) + 2HCl(aq) \rightarrow MgCl_2(aq) + H_2(g)$

   A 0·01
   B 0·02
   C 0·04
   D 0·20

5. The continuous use of large extractor fans greatly reduces the possibility of an explosion in a flour mill. This is mainly because

   A a build up in the concentration of oxygen is prevented
   B local temperature rises are prevented by the movement of the air
   C particles of flour suspended in the air are removed
   D the slow accumulation of carbon monoxide is prevented.

6.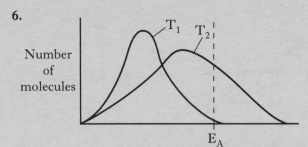

   Kinetic energy of molecules

   Which of the following is the correct interpretation of the above energy distribution diagram for a reaction as the temperature **decreases** from T_2 to T_1?

	Activation energy (E_A)	Number of successful collisions
A	remains the same	increases
B	decreases	decreases
C	decreases	increases
D	remains the same	decreases

7. Which of the following equations illustrates an enthalpy of combustion?

   A $C_2H_6 + 3\tfrac{1}{2}O_2(g)$
   $\downarrow$
   $2CO_2(g) + 3H_2O(\ell)$

   B $C_2H_5OH(\ell) + O_2(g)$
   $\downarrow$
   $CH_3COOH(\ell) + H_2O(\ell)$

   C $CH_3CHO(\ell) + \tfrac{1}{2}O_2(g)$
   $\downarrow$
   $CH_3COOH(\ell)$

   D $CH_4(g) + 1\tfrac{1}{2}O_2(g)$
   $\downarrow$
   $CO(g) + 2H_2O(\ell)$

[Turn over

8. The bond enthalpy of a gaseous diatomic molecule is the energy required to break one mole of the covalent bonds. It is also the energy released in the formation of one mole of the bonds from the atoms involved.

Bond	Bond enthalpy/kJ mol^{-1}
H — H	432
I — I	149
H — I	295

$$H_2(g) + I_2(g) \rightarrow 2HI(g)$$

What is the enthalpy change, in kJ mol^{-1}, for the above reaction?

A +9

B −9

C +286

D −286

9. Which of the following equations represents the first ionisation energy of chlorine?

A $Cl(g) + e^- \rightarrow Cl^-(g)$

B $Cl^+(g) + e^- \rightarrow Cl(g)$

C $Cl(g) \rightarrow Cl^+(g) + e^-$

D $Cl^-(g) \rightarrow Cl(g) + e^-$

10. Which of the following elements has the smallest electronegativity?

A Lithium

B Caesium

C Fluorine

D Iodine

11. A substance melts at 1074 °C and boils at 1740 °C. The passage of an electric current through the molten substance results in electrolysis.

What type of structure is present in the substance?

A Ionic

B Metallic

C Covalent molecular

D Covalent network

12. Which of the following occurs when crude oil is distilled?

A Covalent bonds break and form again.

B Van der Waals' bonds break and form again.

C Covalent bonds break and van der Waals' bonds form.

D Van der Waals' bonds break and covalent bonds form.

13. Which of the following has a covalent molecular structure?

A Radium chloride

B A noble gas

C Silicon dioxide

D A fullerene

14. A metal (melting point 328 °C, density 11·3 g cm^{-3}) was obtained by electrolysis of its molten chloride (melting point 501 °C, density 5·84 g cm^{-3}).

During the electrolysis, how would the metal occur?

A As a solid on the surface of the electrolyte

B As a liquid on the surface of the electrolyte

C As a solid at the bottom of the electrolyte

D As a liquid at the bottom of the electrolyte

15. Which of the following gases contains the smallest number of molecules?

A 100 g fluorine

B 100 g nitrogen

C 100 g oxygen

D 100 g hydrogen

16. Approximately how many atoms will be present in 11·5 litres of carbon monoxide?

(Take the molar volume of carbon monoxide to be 23 litres mol^{-1}.)

A $1·5 \times 10^{23}$

B 3×10^{23}

C 6×10^{23}

D $1·2 \times 10^{24}$

17. $3CuO + 2NH_3 \rightarrow 3Cu + N_2 + 3H_2O$

    What volume of gas, in cm^3, would be obtained by reaction between $100\,cm^3$ of ammonia gas and excess copper(II) oxide?

    (All volumes are measured at atmospheric pressure and $20\,°C$.)

    A 50

    B 100

    C 200

    D 400

18. The following reaction

    $CH_3-CH_2-CH_2-CH_2-CH_2-CH_3 \longrightarrow \bigcirc + 4H_2$

    can take place during

    A dehydration

    B cracking

    C hydrogenation

    D reforming.

19. Biogas is produced under anaerobic conditions by the fermentation of biological materials.

    What is the main constituent of biogas?

    A Butane

    B Ethane

    C Methane

    D Propane

20. Which of the following organic compounds is an isomer of hexanal?

    A 2-Methylbutanal

    B 3-Methylpentan-2-one

    C 2,2-Dimethylbutan-1-ol

    D 3-Ethylpentanal

21. Aspirin is one of the most widely used pain relievers in the world. It has the structure:

    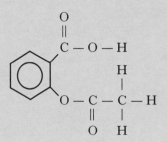

    Which **two** functional groups are present in an aspirin molecule?

    A Aldehyde and ketone

    B Carboxyl and ester

    C Ester and aldehyde

    D Hydroxyl and carboxyl

22. Which of the following hydrocarbons always gives the same product when one of its hydrogen atoms is replaced by a chlorine atom?

    A Hexane

    B Hex-1-ene

    C Cyclohexane

    D Cyclohexene

23. Oxidation of 4-methylpentan-2-ol using copper(II) oxide results in the alcohol

    A losing 2 g per mole

    B gaining 2 g per mole

    C gaining 16 g per mole

    D not changing in mass.

24. Ozone has an important role in the upper atmosphere because it

    A reflects certain CFCs

    B absorbs certain CFCs

    C reflects ultraviolet radiation

    D absorbs ultraviolet radiation.

25. Synthesis gas consists mainly of

    A CH_4 alone

    B CH_4 and CO

    C CO and H_2

    D CH_4, CO and H_2.

26. The following monomers can be used to prepare nylon–6,6.

$$Cl-\overset{O}{\underset{\|}{C}}-(CH_2)_4-\overset{O}{\underset{\|}{C}}-Cl \text{ and } H_2N-(CH_2)_6-NH_2$$

What molecule is released during the polymerisation reaction between these monomers?

A HCl
B H_2O
C NH_3
D HOCl

27. Which of the following statements about nylon and polystyrene is true?

A Both are thermosetting plastics.
B Both are condensation polymers.
C Both give off carbon dioxide and water vapour on burning.
D Both have hydrogen bonds between the polymer chains.

28. Proteins can be denatured under acid conditions.

During this denaturing, the protein molecule

A changes shape
B is dehydrated
C is neutralised
D is polymerised.

29.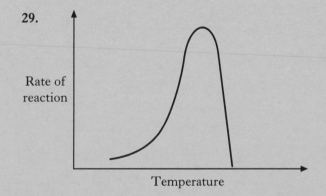

The above diagram could represent

A fermentation of glucose
B neutralisation of an acid by an alkali
C combustion of sucrose
D reaction of a metal with acid.

30. The costs involved in the industrial production of a chemical are made up of fixed costs and variable costs.

Which of the following is most likely to be classified as a variable cost?

A The cost of land rental
B The cost of plant construction
C The cost of labour
D The cost of raw materials

31. Which of the following is produced by a batch process?

A Sulphuric acid from sulphur and oxygen
B Aspirin from salicylic acid
C Iron from iron ore
D Ammonia from nitrogen and hydrogen

32. Consider the reaction pathway shown below.

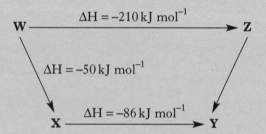

According to Hess's Law, the ΔH value, in kJ mol^{-1}, for reaction **Z** to **Y** is

A +74
B −74
C +346
D −346.

33. Chemical reactions are in a state of dynamic equilibrium only when

A the reaction involves zero enthalpy change
B the concentrations of reactants and products are equal
C the rate of the forward reaction equals that of the backward reaction
D the activation energies of the forward and backward reactions are equal.

34.

$Cl_2(g) + H_2O(\ell) \rightleftharpoons Cl^-(aq) + ClO^-(aq) + 2H^+(aq)$

The addition of which of the following substances would move the above equilibrium to the right?

A Hydrogen

B Hydrogen chloride

C Sodium chloride

D Sodium hydroxide

35. A trout fishery owner added limestone to his loch to combat the effects of acid rain. He managed to raise the pH of the water from 4 to 6.

The concentration of the $H^+(aq)$

A increased by a factor of 2

B increased by a factor of 100

C decreased by a factor of 2

D decreased by a factor of 100.

36. The concentration of $OH^-(aq)$ ions in a solution is 0.1 mol l^{-1}.

What is the pH of the solution?

A 1

B 8

C 13

D 14

37. A white solid dissolves in water, giving an alkaline solution, and reacts with dilute hydrochloric acid, giving off a gas.

(You may wish to refer to the data booklet.)

The solid could be

A copper(II) ethanoate

B potassium carbonate

C ammonium chloride

D lead(II) carbonate.

38. Which of the following is a redox reaction?

A $Zn + 2HCl \rightarrow ZnCl_2 + H_2$

B $NaOH + HCl \rightarrow NaCl + H_2O$

C $NiO + 2HCl \rightarrow NiCl_2 + H_2O$

D $CuCO_3 + 2HCl \rightarrow CuCl_2 + H_2O + CO_2$

39. Phosphorus-32 is made by neutron capture for use as a tracer in phosphate fertilisers.

From which of the following isotopes is phosphorus-32 made by neutron capture?

A $^{31}_{16}S$

B $^{31}_{15}P$

C $^{31}_{14}Si$

D $^{32}_{16}S$

40. The chart below was obtained from an 8-day old sample of an α-emitting radioisotope.

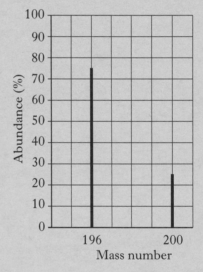

What is the half-life of the radioisotope?

A 2 days

B 4 days

C 8 days

D 12 days

Candidates are reminded that the answer sheet MUST be returned INSIDE the front cover of this answer book.

[Turn over for SECTION B on *Page eight*

SECTION B

1. The structure of a fat molecule is shown below.

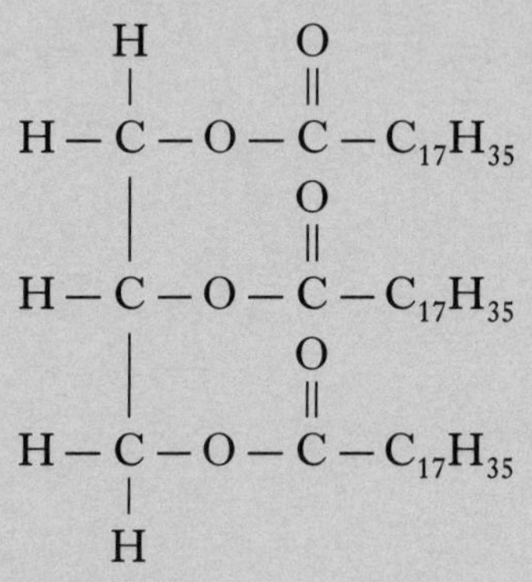

$$\begin{array}{c} H \quad\quad\quad O \\ | \quad\quad\quad\; \| \\ H-C-O-C-C_{17}H_{35} \\ | \quad\quad\quad\; O \\ | \quad\quad\quad\; \| \\ H-C-O-C-C_{17}H_{35} \\ | \quad\quad\quad\; O \\ | \quad\quad\quad\; \| \\ H-C-O-C-C_{17}H_{35} \\ | \\ H \end{array}$$

(a) When the fat is hydrolysed, a fatty acid is obtained.
Name the other product obtained in this reaction.

1

(b) Oils are liquid at room temperature; fats are solid.
Why do oils have lower melting points than fats?

1
(2)

2. The diagram below shows energy changes **A**, **B** and **C** for a reversible reaction.

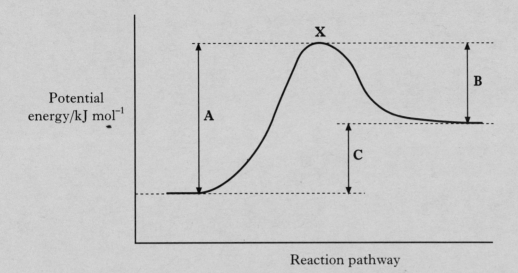

Reaction pathway

(a) What could be used to decrease both **A** and **B** but **not** change **C**?

(b) Give the name of the unstable arrangement of atoms formed at point **X**.

3. (a) Acidified potassium dichromate solution can be used to oxidise some alcohols to aldehydes and then to carboxylic acids, eg

$$\text{ethanol} \longrightarrow \text{ethanal} \longrightarrow \text{ethanoic acid}$$

(i) Name the type of alcohol that can be oxidised to an aldehyde.

(ii) What colour change would be observed when acidified potassium dichromate solution is used to produce ethanoic acid from ethanal?

(b) Ethanol and ethanoic acid react to form the ester, ethyl ethanoate.

ethanol + ethanoic acid ⇌ ethyl ethanoate + water
Mass of one mole   Mass of one mole   Mass of one mole
= 46 g   = 60 g   = 88 g

(i) Clearly describe how the reaction mixture would be heated in the laboratory formation of the ester.

(ii) Use the above information to calculate the percentage yield of ethyl ethanoate if 5·0 g of ethanol produced 5·8 g of ethyl ethanoate on reaction with excess ethanoic acid.

Show your working clearly.

4. Urea, H_2NCONH_2, has several uses, eg as a fertiliser and for the manufacture of some thermosetting plastics.

It is produced in a two-step process.

Step one

$$CO_2(g) + 2NH_3(\ell) \rightleftharpoons H_2NCOONH_4(s) \quad \Delta H \text{ negative}$$
$$\text{ammonium carbamate}$$

Step two

$$H_2NCOONH_4(s) \rightleftharpoons H_2NCONH_2(s) + H_2O(g) \quad \Delta H \text{ positive}$$
$$\text{ammonium carbamate} \qquad \qquad \text{urea}$$

(a) (i) What would happen to the equilibrium position in **step one** if the temperature was increased?

(ii) **Step two** is carried out at low pressure.

Why does lowering the pressure move the equilibrium position to the right?

(b) Draw the full structure for the negative ion in ammonium carbamate.

Marks

1

1

1

(3)

[Turn over

Marks

5. The rate of carbon dioxide production was measured in three laboratory experiments carried out at the same temperature and using excess calcium carbonate.

Experiment	Acid	Calcium carbonate
A	40 cm^3 of 0·10 mol l^{-1} sulphuric acid	1 g lumps
B	40 cm^3 of 0·10 mol l^{-1} sulphuric acid	1 g powder
C	40 cm^3 of 0·10 mol l^{-1} hydrochloric acid	1 g lumps

The curve obtained for Experiment **A** is shown.

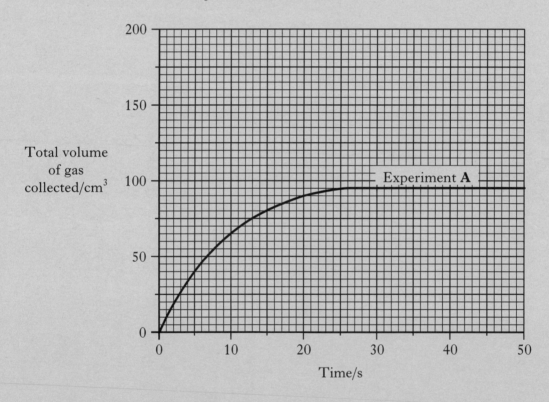

(a) Use the graph to calculate the average reaction rate, in cm^3 s^{-1}, between 10 and 20 s.

1

DO NOT WRITE IN THIS MARGIN

Marks

5. (continued)

(*b*) Draw curves on the graph to show results that could be obtained for experiments **B** and **C**.

Label each curve clearly. 2

(Additional graph paper, if required, can be found on page 31.)

(*c*) Draw a labelled diagram of the assembled apparatus which could be used to carry out this experiment.

2

(5)

[Turn over

6. Uranium ore is converted into uranium(IV) fluoride, UF_4, to produce fuel for nuclear power stations.

(a) In one process, uranium can be extracted from the uranium(IV) fluoride by a redox reaction with magnesium, as follows.

$$2Mg + UF_4 \rightarrow 2MgF_2 + U$$

(i) Give another name for this type of redox reaction.

(ii) Write the ion-electron equation for the reduction reaction that takes place.

(iii) The reaction with magnesium is carried out at a high temperature.
The reaction vessel is filled with argon rather than air.
Suggest a reason for using argon rather than air.

DO NOT WRITE IN THIS MARGIN

Marks

6. (continued)

(*b*) In a second process, the uranium(IV) fluoride is converted into UF_6 as shown.

$$UF_4(s) \quad + \quad F_2(g) \quad \rightarrow \quad UF_6(g)$$

(i) Name the type of bonding in $UF_6(g)$. **1**

(ii) Both UF_4 and UF_6 are radioactive.

How does the half-life of the uranium in UF_4 compare with the half-life of the uranium in UF_6? **1**

(5)

[Turn over

7. The following apparatus can be used to determine the enthalpy of solution of a substance.

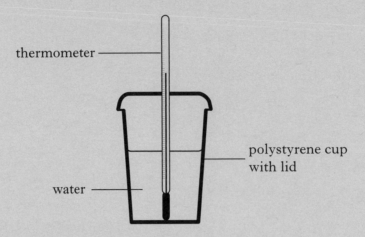

(a) Why was the experiment carried out in a polystyrene cup with a lid?

1

(b) In an experiment to find the enthalpy of solution of potassium hydroxide, KOH, a student added 3·6 g of the solid to the water in the polystyrene cup and measured the temperature rise. From this, it was calculated that the heat energy produced in the reaction was 3·5 kJ.

Use this information to calculate the enthalpy of solution of potassium hydroxide.

Show your working clearly.

2

(3)

8. Acid-base reactions are common in chemistry.

(a) Write the balanced equation for the reaction between copper(II) oxide and nitric acid.

(b) An acid can be thought of as a chemical which can release H^+ ions.

In an acid-base reaction the H^+ ions released by the acid are accepted by the base.

Some acid-base reactions are reversible. In these reactions both forward and reverse reactions involve the transfer of H^+ ions from the acid to the base.

(i) Using the information given above, complete the table showing the acid and base produced when HS^- ions react with H_3O^+ ions.

Acid	Base		Acid	Base
H_2O	NH_3	⇌	NH_4^+	OH^-
H_3O^+	HS^-	⇌		

(ii) Another reversible acid-base reaction is shown.

$$HCO_3^- + OH^- \rightleftharpoons CO_3^{2-} + H_2O$$

In the reverse reaction, state whether the water is acting as an acid or a base.

9. Hydrogen peroxide has a high viscosity.
The structure of hydrogen peroxide is shown below.

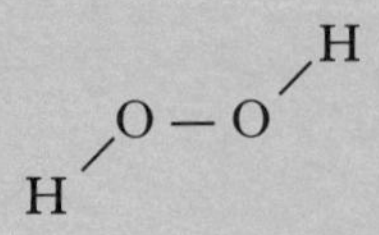

(a) Name the type of intermolecular force that is responsible for hydrogen peroxide's high viscosity.

(b) Hydrogen peroxide may be prepared from its elements.
The equation for the reaction is:

$$H_2(g) + O_2(g) \longrightarrow H_2O_2(\ell)$$

Calculate the enthalpy change, in kJ mol^{-1}, for the above reaction using the enthalpy of combustion of hydrogen from the data booklet and the enthalpy change for the following reaction.

$$H_2O_2(\ell) \longrightarrow \tfrac{1}{2}O_2(g) + H_2O(\ell) \quad \Delta H = -98 \text{ kJ mol}^{-1}$$

Show your working clearly.

9. (continued)

(c) The following solution mixtures were used in a series of experiments involving the reaction between hydrogen peroxide in acid solution and potassium iodide solution.

$$H_2O_2(aq) + 2H^+(aq) + 2I^-(aq) \longrightarrow 2H_2O(\ell) + I_2(aq)$$

	Volume of KI(aq)/cm^3	Volume of H$_2$O(ℓ)/cm^3	Volume of H$_2$O$_2$(aq)/cm^3	Volume of H$_2$SO$_4$(aq)/cm^3	Volume of Na$_2$S$_2$O$_3$(aq)/cm^3	Rate/ s^{-1}
A	25	0	5	10	10	0·020
B	20	5	5	10	10	0·016
C	15	10	5	10	10	0·012
D	10	15	5	10	10	0·008
E	5	20	5	10	10	0·004

(i) From the information in the shaded columns in the table above, what variable is being kept constant throughout the series of experiments?

(ii) What was the aim of the series of experiments?

(iii) Calculate the time, in seconds, for the reaction in Experiment **A**.

10. Poly(ethenol) has an unusual property for a plastic.

(a) What is this unusual property?

(b) (i) A step in the manufacture of poly(ethenol) is shown below.

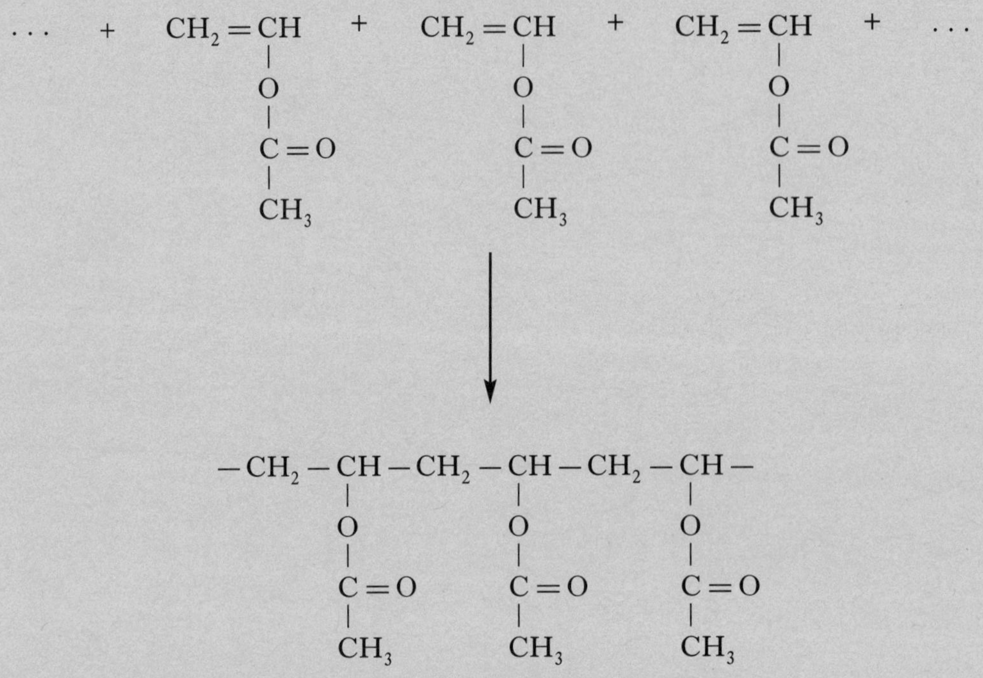

$$\cdots + CH_2=CH\text{-}O\text{-}C(=O)\text{-}CH_3 + CH_2=CH\text{-}O\text{-}C(=O)\text{-}CH_3 + CH_2=CH\text{-}O\text{-}C(=O)\text{-}CH_3 + \cdots$$

$$\downarrow$$

$$-CH_2-CH(O\text{-}C(=O)\text{-}CH_3)-CH_2-CH(O\text{-}C(=O)\text{-}CH_3)-CH_2-CH(O\text{-}C(=O)\text{-}CH_3)-$$

poly(ethenyl ethanoate)

Name the type of polymerisation which takes place in this step.

10. (continued)

(ii) The next step is ester exchange. This involves the removal of the ester side chains by reaction with an alcohol and sodium hydroxide.

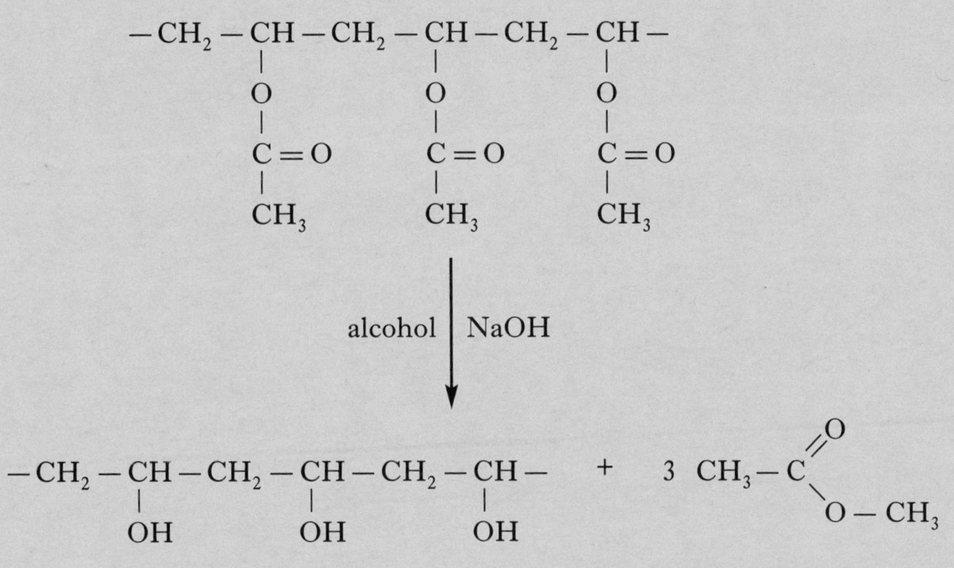

$$-CH_2-CH-CH_2-CH-CH_2-CH-$$
with O–C(=O)–CH₃ ester groups on each CH

alcohol / NaOH ↓

$-CH_2-CH(OH)-CH_2-CH(OH)-CH_2-CH(OH)-$ + 3 $CH_3-C(=O)-O-CH_3$

poly(ethenol)

Name the alcohol used in this step.

1

(3)

[Turn over

11. Respiration provides energy for the body through "combustion" of glucose.

The equation for the enthalpy of combustion of glucose is:

$$C_6H_{12}O_6(s) + 6O_2(g) \longrightarrow 6CO_2(g) + 6H_2O(\ell) \quad \Delta H = -2807 \text{ kJ mol}^{-1}$$

(a) Calculate the volume of oxygen, in litres, required to provide 418 kJ of energy.
(Take the molar volume of oxygen to be 24 litres mol^{-1}.)
Show your working clearly.

(b) In a poisonous atmosphere, a gas mask can be used to provide the oxygen needed for respiration. One type of gas mask contains potassium superoxide, KO_2, which reacts with water vapour to produce oxygen.

The balanced equation for the reaction is:

$$4KO_2(s) + 2H_2O(g) \rightarrow 4KOH(s) + 3O_2(g)$$

(i) Suggest why this reaction allows the same air to be breathed again and again.

(ii) Why is this type of mask also able to remove the carbon dioxide produced by respiration?

12. Ethanoic acid, $CH_3-C{\overset{\displaystyle O}{\underset{\displaystyle OH}{}}}$ (aq), is a weak acid; hydrochloric acid, HCl(aq), is a strong acid.

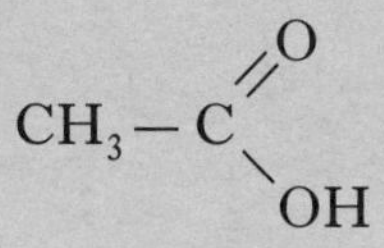

Using ethanoic acid and hydrochloric acid as examples, explain the differences in both pH and conductivity between 0.1 mol l^{-1} solutions of a strong and weak acid.

You may wish to use suitable equations in your answer.

(3)

[Turn over

13. Fuel cells can be used to power cars.

(a) (i) The ion-electron equations for the oxidation and reduction reactions that take place in a methanol fuel cell are:

$$CH_3OH(\ell) + H_2O(\ell) \rightarrow CO_2(g) + 6H^+(aq) + 6e^-$$

$$3O_2(g) + 12H^+(aq) + 12e^- \rightarrow 6H_2O(\ell)$$

Combine the two ion-electron equations to give the equation for the overall redox reaction.

(ii) The equation for the overall redox reaction in a hydrogen fuel cell is

$$2H_2(g) + O_2(g) \rightarrow 2H_2O(\ell)$$

Give a disadvantage of the methanol fuel cell reaction compared to the hydrogen fuel cell reaction.

DO NOT WRITE IN THIS MARGIN

Marks

13. (continued)

(*b*) The hydrogen gas for use in fuel cells can be produced by the electrolysis of water.

Hydrogen is produced at the negative electrode as shown.

$$2H_2O(\ell) \quad + \quad 2e^- \quad \rightarrow \quad H_2(g) \quad + \quad 2OH^-(aq)$$

Calculate the volume of hydrogen gas produced when a steady current of 0·50 A is passed through water for 30 minutes.

(Take the molar volume of hydrogen to be 24 litres mol^{-1}.)

Show your working clearly.

3

(5)

[Turn over

14. The structure of an ionic compound consists of a giant lattice of oppositely charged ions. The arrangement of ions is determined mainly by the "radius ratio" of the ions involved.

$$\text{radius ratio} = \frac{\text{radius of positive ion}}{\text{radius of negative ion}}$$

The arrangements for caesium chloride, CsCl, and sodium chloride, NaCl, are shown below.

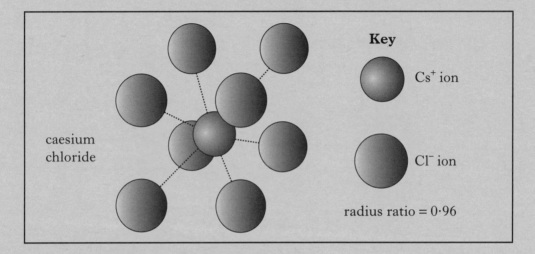

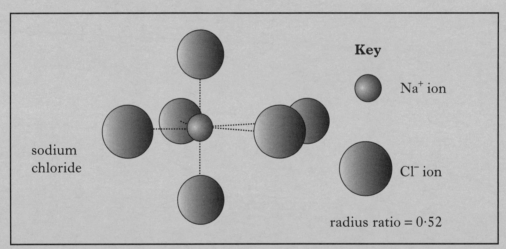

(a) By using the table of ionic radii on page 16 of the data booklet, calculate the radius ratio for magnesium oxide, MgO, and state which of the two arrangements, caesium chloride or sodium chloride, it is more likely to adopt.

DO NOT WRITE IN THIS MARGIN

Marks

14. (continued)

(*b*) The enthalpy of lattice breaking is the energy required to completely separate the ions from one mole of an ionic solid.

The table shows the enthalpies of lattice breaking, in kJ mol^{-1}, for some alkali metal halides.

Ions	F^-	Cl^-	Br^-
Li^+	1030	834	788
Na^+	910	769	732
K^+	808	701	671

Write a general statement linking the enthalpy of lattice breaking to ion size. **1**

(2)

[Turn over

15. Vitamin C is required by our bodies for producing the protein, collagen. Collagen can form sheets that support skin and internal organs.

(a) (i) There are two main types of protein.
Which of the two main types is collagen?

1

(ii) Part of the structure of collagen is shown.

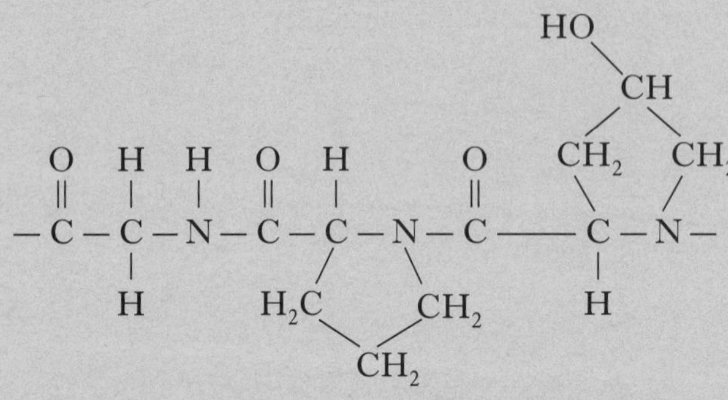

[Structure showing: —C(=O)—C(H)(H)—N(H)—C(—CH₂—CH₂—CH₂ ring)—C(H)—N(H)—C(=O)—C(—CH₂—CH(OH)—CH₂ ring)—N(H)—]

Draw a structural formula for an amino acid that could be obtained by hydrolysing this part of the collagen.

1

15. (continued)

(b) A standard solution of iodine can be used to determine the mass of vitamin C in orange juice.

Iodine reacts with vitamin C as shown by the following equation.

$$C_6H_8O_6(aq) + I_2(aq) \rightarrow C_6H_6O_6(aq) + 2H^+(aq) + 2I^-(aq)$$
vitamin C

In an investigation using a carton containing $500 \, cm^3$ of orange juice, separate $50.0 \, cm^3$ samples were measured out. Each sample was then titrated with a $0.0050 \, mol \, l^{-1}$ solution of iodine.

(i) Why would starch solution be added to each $50.0 \, cm^3$ sample of orange juice before titrating against iodine solution?

1

(ii) Titrating the whole carton of orange juice would require large volumes of iodine solution.

Apart from this disadvantage give another reason for titrating several smaller samples of orange juice.

1

(iii) An average of $21.4 \, cm^3$ of the iodine solution was required for the complete reaction with the vitamin C in $50.0 \, cm^3$ of orange juice.

Use this result to calculate the mass of vitamin C, in grams, in the $500 \, cm^3$ carton of orange juice.

Show your working clearly.

2

(6)

DO NOT WRITE IN THIS MARGIN

Marks

16. Carbon compounds take part in many different types of reactions.

ethanol ⟶ ethene

ethyne $\xrightarrow{\text{addition}}$ compound **Y** $\xrightarrow{\text{hydrogenation}}$ monochloroethane

cyclohexanol $\xrightarrow{\text{oxidation}}$ compound **Z**

(*a*) Name the type of reaction that takes place in the formation of ethene from ethanol. **1**

(*b*) Draw a structural formula for

(i) compound **Y**; **1**

(ii) compound **Z**. **1**

(3)

[*END OF QUESTION PAPER*]

ADDITIONAL SPACE FOR ANSWERS

ADDITIONAL GRAPH PAPER FOR QUESTION 5(*b*)

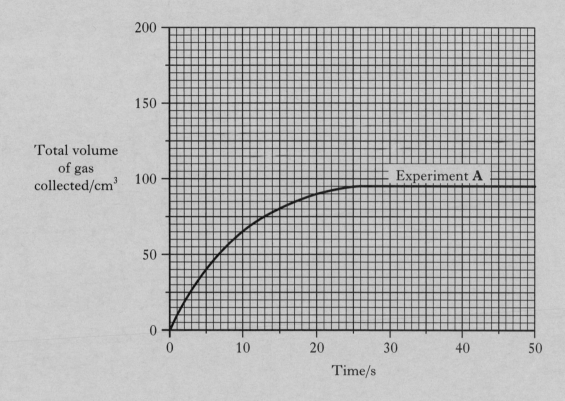

Acknowledgements

Leckie & Leckie is grateful to the copyright holders, as credited, for permission to use their material.

The following companies/individuals have very generously given permission to reproduce their copyright material free of charge:
Michael J. Pelczar for a diagram from *Microbiology, Concepts & Applications* by Pelczar, Chan & Kreig (2002 paper p 18).